ESSAYS FOR THE THIRD CENTURY

Vincent Davis, General Editor

ACCESS TO ENERGY

2000 and After

MELVIN A. CONANT

THE UNIVERSITY PRESS OF KENTUCKY

Library of Congress Cataloging in Publication Data

Conant, Melvin.
    Access to energy, 2000 and after.

    (Essays for the third century)
    Includes bibliographical references and index.
    1. Petroleum industry and trade.   2. Power
resources.   3. Energy policy.   4. Economic forecasting.
I. Title.   II. Series.
HD9560.5.C58   333.7   79-15015
ISBN 0-8131-0401-7

Scholarly publisher for the Commonwealth
serving Berea College, Centre College of Kentucky,
Eastern Kentucky University, The Filson Club,
Georgetown College, Kentucky Historical Society,
Kentucky State University, Morehead State University,
Murray State University, Northern Kentucky University,
Transylvania University, University of Kentucky,
University of Louisville, and Western Kentucky University.

Editorial and Sales Offices: Lexington, Kentucky 40506

TO CHRISTA

Contents

Abbreviations

BP	British Petroleum
CFP	Compagnie Française des Petroles
CIEC	Conference on International Economic Cooperation
IAEA	International Atomic Energy Agency
IEA	International Energy Agency
IEI	International Energy Institute
IFAD	International Fund for Agricultural Development
INOC	Iraq National Oil Company
IOC	International Oil Companies
IPC	Iraq Petroleum Company
IRB	International Resources Bank
LDC	Less Developed Country
LNG	Liquefied Natural Gas
MMB/D	Million Barrels per Day
MMB/DOE	Million Barrels per Day Oil Equivalent
NIEO	New International Economic Order
NIOC	National Iranian Oil Company
OAPEC	Organization of Arab Petroleum Exporting Countries
OECD	Organization for Economic Cooperation and Development
OPEC	Organization of Petroleum Exporting Countries
Socal	Standard Oil of California
TPC	Turkish Petroleum Company
UNCTAD	United Nations Conference on Trade and Development

Preface

IN THE last fifty years we have experienced the greatest shift in the source of energy ever known, comparable perhaps only to the discovery of fire. In the first half of the twentieth century, coal maintained its preeminent position as the major fuel source for the industrialized world. The discovery of large amounts of oil in the Soviet Union, the United States, and, above all, the Middle East, along with the comparative ease of oil extraction, its extraordinary range of uses, and its easy transportation, led to profound change in fuel sources. During the 1960s oil replaced coal as the primary energy source for the industrialized world. This shift accelerated industrial development worldwide, thus further increasing the demand for oil. (Yet, despite the magnitude of the increase in demand for oil, which rose from 3 million barrels a day in 1925 to 60 million in 1977, nearly half of the world's energy consumption is derived from wood, peat, and dung, the so-called noncommercial fuels.)

Oil is a finite resource; we can only guess when its supply will be exhausted. We know only that, for over a decade, our discovery or finding rate for very large reserves has been virtually zero. In short, the world has been living off its oil capital. The cherished reserves-production ratio of 10:1 or even 15:1 is no longer the general condition outside the Middle East.

It is a basic theme of this book that the fundamentals of our energy situation will not greatly change for many years. Events such as the shift in Iran, which occurred too late to be fully incorporated in the text, highlight the long-range picture of an increasingly tight supply of oil in world trade. The surplus of oil referred to in the discussion disappeared with the decrease in Iranian production.

The prudent citizen or government official or business executive must plan and work his way through what is left of the petroleum age. Sometime in the next century he will have other sources of energy. Between now and then, however, access to energy placed in world trade, which meets the deficiencies of most states, is bound to become an enduring interest of governments. Competition for available supply is likely to be intense and war over access to vitally needed supplies is by no means impossible.

In this volume I attempt to trace the adequacy and continuity of energy supply in world trade. To help understand what may lie ahead, we must be aware of the past; thus, much of this book is intended to inform the reader of factors that affect the relations among states in the context of their energy requirements.

In this discussion of possible issues which access to energy will pose for importing nations, I have been helped especially by Charles K. Ebinger in the preparation of this volume, particularly the chapter on nuclear energy, and by Christa G. Conant, who contributed the chapter on ocean energy sources. Ms. Jeanne-Marie S. Peterson has assisted me in the preparation of this volume. My thanks to each, and to the University Press of Kentucky for giving me the opportunity to make what are summary judgments in hopes some perspective can be given to a very complex question: how we and others can meet our needs for imported energy supply until new technologies eliminate it from our list of concerns.

Introduction

OIL is featured in this volume because of its prominence among the
primary energy fuels and because of its unequaled importance in
world energy trade. It is emphasized also because the history of the
rise of international oil to its present position of importance offers
insight into both current access and other energy commodities of
potential significance. Oil has been transformed from a commercial
commodity traded by private international oil companies (the
"Seven Sisters"*) into a strategic raw material, the access to which
is now determined by sovereign governments; the same political
factors presently involved in access to oil may also be embedded in
the obtaining of nuclear/uranium supply.

From the beginning of the history of foreign oil, governments
have sponsored their oil companies in the obtaining and retention
of concession agreements. These concessions were originally obtained from the nominal governments of the region, but the overwhelming supremacy of British imperial interests, and those of
France, firmly established oil as a strategic commodity of great
national importance under western control. A variety of de-
pendencies developed during the end of the nineteenth century, and
these became particularly important following World War I and

*United States companies: Standard Oil of New Jersey (now Exxon), Mobil,
Texaco, Gulf Oil, and Standard Oil of California. British companies: British Petro-
leum (majority control by the British government) and the British-Dutch group:
Royal Dutch/Shell.

the League of Nations' mandate system, which reaffirmed and extended the supremacy of western empire over the political and economic fortunes of the colonial world, especially the Middle East.

With the collapse of empires after World War II and the rise of political and economic nationalism among the former colonies, access to oil became a crucial consideration to importing nations as the customary oil-concession system began to crumble. Virtually every major supplier of oil into world trade had a colonial background; with the spread of independence, the politics of oil now assumed vital importance in setting the terms on which it would be supplied.

By the 1960s, most of the industrial and developing world had come to depend upon imported oil, in ever-increasing volumes, and this dependence led to economic and political confrontation. The suppliers of oil, those former colonies, were not about to forget—or to allow others to forget—what they regarded as exploitation by the West. The consequences on world politics and the economics of these political concerns about oil supply have only begun to be studied. We can be assured that questions of access to energy will be with us until the next generation of energy sources—fusion, solar, the oceans, and the deep heat of the earth itself—are technologically available in sufficient quantity to meet world energy needs, not until well into the twenty-first century.

Our national requirement is to conserve present energy resources and to develop additional and alternative sources as rapidly as possible. An energy policy that disciplines our use of nonrenewable energy sources—the fossil fuels of coal, gas, and oil—is fundamental. In a complementary effort we must use every means to incorporate the interests of those with raw materials into the world economic and financial system, created by the users of those commodities, to help ensure continuity and adequacy of supply.

Failure to develop all our alternative energy resources will lead to an ever-burgeoning rise in demand on the international petroleum market. Already the Organization for Economic Cooperation and Development (OECD) estimates that, if current policies governing supply expansion and conservation are continued, the oil import demand of the OECD nations by 1985 could amount to 35 million barrels per day (MMB/D), up 10 MMB/D from 1974. When the rest of the noncommunist world is included, the level of oil import demand could reach about 39 MMB/D.[1] Other analyses predict that world oil import levels may reach as high as 47-51 MMB/D by 1985.[2]

There are grave financial implications in this general dependence on imported oil. If we take the oil import levels projected by the OECD, the cost of oil imports to the noncommunist world could reach $200 billion in 1985 and nearly $300 billion in 1990 (in current dollars). The financial implications of this increase are all the more awesome when seen against today's serious and mounting problems in attempting to cope with the present deficits of some industrial states and many developing countries. While the non-oil-developing countries are not included in these numbers, their oil import bills assume staggering proportions and the difficulties in their being able to continue to meet them are readily apparent. There is a growing likelihood that the $50 billion of loans that American banks have made to the non-oil less-developed countries (LDCs) may never be repaid; a large part of these financial obligations have their origin in the increased price of oil, as set by the Organization of Petroleum Exporting Countries (OPEC). Indeed, the debt-burden of the LDCs is assuming such vast proportions that there are fears that many are approaching the limits of their creditworthiness and will find it increasingly difficult and perhaps impossible to meet past obligations.

Non-OPEC countries may have to borrow at least $40 billion per year just to stay afloat financially. This accumulation, year after year, could create such huge debts that most LDCs and even most western industrial nations (possibly even the United States itself) will be confronted by repayment obligations which could prove unmanageable. Most oil-exporting nations see this situation, however, as one in which they are obtaining, for the first time, only their rightful share of the proceeds from their resource.

If we are to consider the complicated issue of energy access, we must understand the startling changes that have occurred in the international energy environment. These include the extremely rapid rise in energy consumption throughout the world; the changing patterns of ownership of oil concessions from those of the international oil companies of the consuming world to those of the governments of the oil-producing states; the continued importance of the political and economic legacy of the colonial experiences of all the key oil-producing states; and the growing intervention by governments (consuming and producing) in the international energy marketplace.

In this review, I will examine the nature and magnitude of the energy problems that will continue to confront the world over the next twenty-five years (and thus into the next century); I intend to stress how questions of access to vital energy supplies will dominate our geopolitical environment for as long as most states must rely upon foreign sources. And I will emphasize the importance of contemporary efforts to improve upon the general terms affecting commodity trade as exemplified in North-South dialogues—a debate that includes energy.

I will look then at the experiences of several key producing and consuming states to illustrate how political nationalism influences their perspectives and how these differences will affect access to their energy. The various multilateral initiatives under way to see us through the transition period are then considered and appraised, for these efforts will profoundly shape our energy future.

Against this background I then discuss the potential for cooperation and/or confrontation over access to energy during the waning years of the century. In this regard, I shall examine the potential impact that developments on the "frontiers of energy" may have in reducing the potential for confrontation. I shall argue that the timely development of all energy resources is critical as each of the new forms—fusion, radiation[3]—is domestic; their availability on a sufficient scale could signal the beginning of the end for the geopolitical aspects of energy and a waning of the competition among states for their supply. I will concentrate overall on the needs of the industrial world, for these are literally overwhelming. The developing countries, of course, will not escape the petroleum age, but their inability to compete for available supplies almost requires some form of allocation or special terms; and even so, of the nearly 30 MMB/D of oil in world trade, only some 2-3 MMB/D go to the developing world.

Finally, I will question the continued viability of the concept of sovereignty over natural resources in an increasingly inter-dependent world. We may need a different concept, one which holds that some raw materials are, in fact, a global commodity— access to which may move beyond contemporary conceptions of international law and the practices of states. In doing so, I will raise the question most central to this entire analysis, namely, whether or not a state or group of states has a right to withhold supplies (energy, food) which have become vital to others or whether some concept of international supervision over production may yet be necessary, however difficult to accomplish. From these consid-erations it is possible to look even further ahead at those aspects of energy that may confront coming generations, many of which have their origins in our own time.

1

The Nature of the Crisis
Energy Supply/Demand, 1977-2000

BECAUSE of accidents of geology, the greater part of oil discovered and produced now lies outside the sovereignty of the industrial nations. What makes access to oil so critical a concern is the general dependence of most nations upon *imported* oil; with comparatively few exceptions, even those states possessing substantial oil reserves of their own (the United States, Great Britain, Norway, and perhaps Canada and the USSR in the years ahead) will be vitally dependent on oil placed in international trade by the producing-exporting countries chiefly in the Middle East. It is unlikely that the prominence of the region as the principal supplier of oil in world trade will be significantly diminished within this century.

It will be argued that continued dependence upon imported oil, and perhaps uranium, for a great many years ahead is avoidable for the United States, at least if sufficient and timely efforts were put into the exploitation of shale, coal gasification, nuclear, solar, geothermal, and other domestic sources. Nevertheless, the extraordinary lead times required to move from the laboratory to the consumer, the seemingly limited transportability of such a source as geothermal energy, and environmental and safety considerations place the displacement of oil as a key energy source well into the next century.

In effect, the industrial and developing nations are now in a transition period, moving out of the petroleum era into the next energy revolution—most likely to reflect a renaissance of coal in conventional, liquefied, and gasified form, in nuclear power, and, increasingly, in the use of various forms of radiation. While a few people will argue that these more exotic sources of energy will be available soon and in adequate amounts to permit the developing nations to escape the petroleum age, most energy observers believe that this is not the case. Partly because of the availability of oil and partly because of the length of time required to develop alternative energy sources, it appears that oil will continue to be the primary energy source for a number of decades. The need for urgency in developing these alternative sources is self-evident; oil is not an infinite resource, but for the next decade and longer it will be the most essential one.

Where energy is provided by a wholly domestic resource, the probability of international contention over its exploitation seems remote, compared to the extreme tension that can develop over access to energy resources owned by another nation or where, as in certain forms of energy derived from the oceans, international disputes over resource rights exist.

In discussing current and future access to energy, I am primarily concerned with the present major source of energy—oil— and the potentially future major source—uranium. It cannot be said, however, in focusing attention on these two kinds of fuel, that I am ignoring the contributions which may yet come from alternative sources. In fact, as a way to exemplify many of the obstacles and problems facing the development of some of the exotic sources of energy, one chapter is devoted to the ocean and its various possible energy resources. Obviously, far less is known about ocean energy than about oil; it is, however, a potentially enormous source. The efficient development of ocean-derived energy will be affected by many factors of which we should be aware, such as technological sophistication, overall and relative costs, international disputes caused by questions of ownership, and lead times involved, so that we might be better able to deal with them at the proper time. Today, however, and probably into the next century, the continuous supply of oil is of vital interest. Hence, we begin our look into the future by being certain we know enough of the past to understand the politics of oil and how some of the same causes may help determine the ways in which uranium can be acquired.

PERSPECTIVE ON WORLD ENERGY

Although the 1950s and 1960s had witnessed the birth of many new states, as old colonial empires gave way before the forces of modern nationalism, trading relationships between Great Britain and France and their former colonies remained largely intact. At the same time, Japan and Germany embarked on ambitious export-promotion drives throughout the world to ensure their access to vital raw materials. The United States, with its abundant reserves of raw materials and its ever-expanding foreign trade, was placed in a unique position among the major industrialized powers in the postwar world.

Energy forecasters had little reason to be concerned about the adequacy of supply. Given the magnitude of the problems they confronted, they did not anticipate that economic growth would be so swift as to cause rapid increases in energy demand. Still, one reason for the degree of complacency about energy which existed during this time lay in the fact that forecasts of energy growth are customarily underestimated; along with this, there has also always been enough surplus supply to compensate for the unexpected growth rates.[1] By the late 1960s and early 1970s, however, the simultaneous burgeoning rise in energy consumption in all the major nations of the industrialized world, and among the developing states as well, made it increasingly apparent that competition for access to foreign markets and raw material supplies— now including energy sources—was beginning to cause strains in traditional trading relationships and even within the western alliances.

During the postwar period, 1945-1973, the industrial expansion of Europe, Japan, and the United States depended on their continued access to cheap energy reserves throughout the world. Although the western world's dependence on oil supplies rose dramatically from 1960 onwards, there was a singular lack of awareness on the part of many of the industrialized countries of the growing changes in the control of access to vital supplies. To understand how this strategic vulnerability evolved, if not the political shortsightedness which accompanied it, it is necessary to examine the rapid growth in energy requirements and the increase in demand for imported energy that occurred after 1960.

Between 1950 and 1977, energy consumption nearly tripled in the United States, Europe, and Japan. At the same time, the importance of oil to the energy economy of the United States rose from 20 to 35 trillion BTUs while its import dependence rose from 23 percent to 46 percent. It was during this period that the United States ceased to be able to meet its energy requirements from indigenous resources and ceased to be the emergency oil supplier to Japan and its North Atlantic Treaty Organization allies. While the level of American import dependence grew, Europe's and Japan's almost total dependence on imported oil also continued, and the worldwide volumetric demand for oil grew tremendously. In 1950 the world consumed 11 MMB/D; in 1978 it required nearly 60 MMB/D. (In 1950 coal was at an oil equivalent of about 22 MMB/D. In spite of increasing demand for coal—35 MMB/D oil equivalent in 1978—coal had lost out to oil and to gas as well, which in 1950 was the equivalent of less than 2 MMB/DOE growing to 26 MMB/DOE in 1978.)

During the 1960-1976 period, the Middle East and Africa became the most important sources of oil for Western Europe and Japan, while their importance to the United States nearly tripled. Of all the major industrialized nations, only the USSR was (and remains even today) energy self-sufficient during this period,[2] a matter of considerable strategic importance when contrasted with the rise in United States dependence on oil imports.

Although organizations, such as the Club of Rome, have warned about impending shortages of vital raw materials, it took the dramatic events surrounding the OAPEC (Organization of Arab Petroleum Exporting Countries) embargo of October 1973—itself a consequence of the Arab-Israeli October war—to demonstrate to the industrialized world the degree of its economic and political vulnerability to a denial of energy supplies from the Middle East, especially from the Persian Gulf. Throughout the period of spiraling dependence on imported oil, the industrial world remained largely oblivious to its growing dependence on foreign sources. By the mid-1960s, France, in the OECD Energy Committee, drew sharpest attention to the consequences likely to follow. But it was not until the winter of 1973-1974, with the embargo and fourfold price increases, that the question of access to raw materials in general and to energy in particular had become of paramount importance to the industrialized world.

In the changing international atmosphere resulting from the embargo, it became increasingly apparent that continued access to raw materials by the industrialized countries would no longer be governed by traditional colonial-style arrangements or alliance structures. Rather, economic and political factors outside the industrial countries' exclusive control would have to be considered. Likewise, in this changing geopolitical environment, it became evident that cheap energy was a thing of the past and that in the future the question of security of access or continuity of flow would be at least as important as the question of price. Until 1973 the degree of vulnerability of the industrialized world to a denial of cheap raw material supplies from the third world was seldom viewed as a security problem by most strategic analysts in the industrialized world.

We know now that competition for energy and other vital raw material supplies will play an increasingly important role in the economic development of nations and in the shifting balance of world power from one dominated by the consumers of raw materials to one in which the producers have additional influence.

THE WORLD ENERGY SITUATION AFTER 1977

As of late 1978, the world confronted a situation which warned that through this century the economic well-being of the developed world and the non-oil less-developed countries will be directly dependent on OPEC (particularly the Arab members') oil reserves. Even vigorous measures taken in conservation and in alternative resource development will not fundamentally reduce dependence on imported oil for the rest of this century. Given the long lead times for exploration and development of alternative energy resources, the availability of energy supplies through much of the 1980s has already been determined. Positive decisions and commitments made in 1979 will not be reflected in any nation's overall domestic energy supply/demand balance until after 1985. If these decisions are not made, the world could face a severe energy crisis in the 1980s as world oil demand rises to between 67 and 73 MMB/D by 1985 (given the improved general economic growth rate which is the precondition to social progress). In order to meet these demand levels, OPEC, which produced around 30.9 MMB/D in 1976, will have to produce somewhere between 42.8 and 51.2 MMB/D.[3] The United States is deeply involved in these events as the single largest importer and consumer of oil.

The United States' growing dependence on imported oil is further complicated by the fact that although most energy policymakers agree on the need for an energy policy, there is little agreement on what that policy should be. The nature of this debate centers not only on the question of whether one particular nonrenewable source should be developed over another, e.g., coal versus nuclear versus oil, but also on the question of whether the continuation of current policy, with its heavy reliance on fossil fuels, is indeed the proper route to follow; many will argue that there is no ready alternative.

Increasingly, critics such as Amory Lovins argue that rather than expanding centralized high-energy technologies (oil refineries, LNG [liquefied natural gas] processing facilities, nuclear power plants, and enrichment facilities), which are astronomical in cost and relatively energy-inefficient, the United States should begin to make a serious commitment to the efficient use of energy, namely, a rapid, intensive effort toward the development of renewable energy sources matched in scale and in energy quality to end-use needs, while relying necessarily on fossil fuel technologies during the transition period.[4]

While the merits of these widely different strategies for the long-term future lie beyond the focus of this inquiry, one critical factor remains, namely, that during the transition period (1979-2000) of the conversion away from oil, gas, and coal to solar, nuclear, wind, geothermal, and other sources, access to foreign energy—oil and uranium—will remain a pivotal ingredient in the foreign policy of all the countries in the world community.

This factor, more than any other, should give cause for alarm; as the world moves into the last quarter of the twentieth century, it is heading inexorably toward a major new energy crisis. By 1985 or, at the latest, 1990, there will be an endemic imbalance between the availability of oil, gas, and nuclear supplies and world demand. Even by this time, potentially new rich petroleum areas, such as the Svalbard Shelf, Antarctic, the Argentine Continental Shelf, and the Soviet Arctic, will not be available in sufficient quantities fundamentally to affect the world energy situation. Similarly, even by 1985-1990, it is unlikely that the economics of coal synthetics ($20-25/barrel oil equivalent at 1977 market prices) will have allowed these fuels to make sizable inroads into world energy usage.

Although a dramatic rise in the level of international coal trade could help to alleviate serious supply shortages in oil, gas, and nuclear, in the short term, environmental constraints, logistical difficulties, and lack of a labor force in the major coal-producing nations evoke doubt that coal in international trade will exert much influence in ameliorating the supply/demand pressures on oil and gas supplies. About one-third of the coal in world trade today originates in the United States, constituting (in 1976) about 10 percent of its annual production. Europe does not think it can add significantly to its current production and believes it must look mainly to the United States for any large increase—an increase that is unlikely to exceed present export levels (70 million tons) over the next decade as increased United States production is destined for domestic use—and perhaps to reduce its dependence upon imported oil rather than to meet the needs of allies.[5]

COMPETITION OVER RESOURCES

The result of this situation will be growing competition among the industrialized nations for ever-dwindling petroleum supplies. In such a scenario, the geopolitics of access to energy will assume a pivotal role in intra-OECD, East-West, and North-South relations.

The present world oil surplus is an anomaly, real for the moment but illusory in the longer run, arising from reduced recession-caused levels of industrial activity in the developed world, the availability of large volumes of heavy oil in Alaska and the North Sea, and the continued willingness of Saudi Arabia to produce oil in excess of present revenue needs.

As the current surplus of world oil recedes, the race to sell or barter arms and/or industrial technology (including nuclear) for guaranteed access to energy supplies will grow with potentially disastrous consequences for the stability of the world. There are already warning signals in the continuing division between the United States, on the one hand, and Europe and Japan, on the other, on the need to develop the breeder reactor and enrichment and reprocessing facilities; in continuing differences among the industrialized countries over North-South issues; in France's African diplomacy (as seen in the intervention in Chad and Zaire), designed to keep access to vital uranium and raw materials; and in Norway's determination to control the development of its North Sea resources which presently means limiting production to meet its own revenue needs and no more.

In addition, the Europeans and Japanese have always been skeptical as to whether the United States would share vital energy supplies in the event of a crisis, a concern that has been heightened by President Carter's stance on the breeder reactor, the 1975 curtailment of enriched uranium sales, and most recently the decision in July 1977 specifically not to sell Alaskan oil to Japan. Furthermore, while the emergency-sharing formula agreed to in the International Energy Agency (IEA) commits the United States to help Europe and/or Japan in the event of an energy shortfall, the Europeans and Japanese are quite aware that United States support of Israel in the wake of another Middle Eastern war is the most likely event that could precipitate another embargo.

From the European/Japanese perspective, it is unlikely that they would be the target of an Arab embargo in the event of a crisis unless they facilitated American military resupply of Israel. Because of this perception, major allies of the United States believe that the IEA emergency oil-sharing mechanism was initiated by the

United States to make sure that the OECD nations, in the event of a crisis, would speak with one voice and remain united vis-à-vis the OPEC oil producers, thus protecting larger American strategic interests. In turn, the United States believed that unless it assured Europe and Japan of its commitment to their security, the potential for a division in OECD ranks could not be averted in the event of another Middle East crisis.[6] Nonetheless, the Europeans remained skeptical of American policy in the IEA, particularly when in the middle of the negotiations to establish the IEA, the United States announced the creation of a "special relationship" with Saudi Arabia from which the others were pointedly excluded.

While the IEA emergency oil-sharing mechanism may help to reduce the level of industrial-country competition over access to oil in the event of another embargo, energy forecasts of government, industry, and academia often neglect to consider that the coming energy crisis could occur not only as a result of an absolute shortage in any of the supply/demand balances of the various nonrenewable fuels but also as a result of inadequate quantities of alternative energy resources to relieve the supply/demand gap.

Because of a host of technical, economic, and political considerations, future production by the major oil-producing countries may not necessarily prove to be the same as their future capacity to produce. For example, although it is projected that, by 1985, Saudi Arabia could produce 20 MMB/D, it may not be in its national interest to produce at that rate if its revenue needs (current account) could be met at a production level of less than half that volume; furthermore, production at the level of 20 MMB/D would deplete the Saudis' oil resources at an excessive and highly imprudent rate. A Saudi decision to produce at the lower level would have profound economic and political ramifications on the geopolitics of energy. To the extent that the Saudis curtailed production to this level, the United States could find itself in direct competition with its major allies, and possibly the Soviet Union, for access to the limited oil supplies then in the world market. Such competition would generate acute tensions leading to corrosive competition, intraalliance fissures, and increased East-West confrontation. If some European nations or Japan saw their long-term energy interests more in concert with the potentially resource-rich Soviet Union (gas, coal, oil, uranium), new political relationships could emerge that would fundamentally alter the traditional pattern of postwar economic and security relationships.

Similarly, while substantial increases in production are projected for new producing areas such as Mexico, there is danger in relying upon future production estimates on the basis of present geologic information. It is even less prudent to estimate what areas such as the Arctic latitudes might supply; in any event, it is highly unlikely that any discoveries over the next several decades will result in significantly reduced dependence upon the Middle East. Furthermore, while forecasts of potential supply will continue to be made, they offer no insight into the decision-making process in oil-producing countries where oil production levels may be governed by domestic political and economic decisions, not only by the level of world oil demand. Theoretical concepts of "market forces" balancing demand and supply are no longer determinants (if they ever were for oil); the cost is worldwide economic and social chaos.

The existence of proved oil reserves is only an indication of what volumes may be produced. Too often, energy analysts point to the positive relationship between reserves and production as the best indicator of future supply levels when in fact the rate of production may be determined by factors exogenous to the level of world energy demand. What will determine the future level of oil production will be the oil producers' own social, political, and economic development goals as partially reflected in their development plans and their international ambitions.[7] In some cases, these goals may imply production levels at considerable variance with the import requirements of the OECD nations. When this possibility is viewed in conjunction with the ability of some politically mobilized interest groups (environmentalists and those with safety concerns) within the industrialized countries to delay resource-development projects through protracted legal suits and physical disruption (Seabrook), it should be apparent that the world energy supply/demand imbalance can be adversely affected by nonmarket forces.[8]

Let us examine the case of Saudi Arabia, which produced close to 9 MMB/D in 1977, although it is estimated that all its development goals could be achieved at a current production level of 5 MMB/D. If the Saudis in December 1976, on the eve of the OPEC conference in Doha, Qatar, had decided to drop their production to 3 MMB/D, for which sound economic arguments could have been mustered, OPEC's estimated spare potential of 7 MMB/D would have evaporated.

If several other resource-rich OPEC countries without the absorptive capacity to use their excess oil revenues had acted in concert with Saudi Arabia, the effect on oil prices in 1977, compounded by a then projected 6.5 percent increase in demand, could have been staggering.

One irony of the situation is that the domestic energy policies of the OECD countries contribute another vital ingredient to the geopolitics of energy. The OECD nations have yet to face up to the fact that their failure both to curtail consumption and to undertake energy development programs in the 1978-1985 period could contribute as significantly to the creation of an energy crisis after 1985 as could the actions of the oil exporters.[9]

To demonstrate how the failure of the OECD nations to take such measures will affect the demand for oil, one has only to examine the estimates for free world oil consumption in 1975, 1980, 1985, and 1990. Whereas in 1975 the free world consumed about 44.2 MMB/D, this could rise to 54.8 MMB/D in 1980, 66.9 MMB/D by 1985, and 76.2 MMB/D by 1990.[10] Demand levels of this magnitude will exert tremendous upward pressures on price as world supply falls short of world demand. In such a scenario, OPEC's power can rise substantially, thus allowing OPEC to levy major increases in the world price of oil. Nonetheless, one critical factor remains: a wishful policy based on the expectation of an oil price break or on vast new additions from hypothetical reserves is both dangerous and irresponsible. Unless this is understood, the critical issues of access to energy during the remainder of the twentieth century cannot be fully appreciated and will certainly not be met in a timely manner.

With all these different factors working upon it, the international energy environment is likely to undergo numerous changes throughout the rest of the century. It will probably move toward a wider definition and dispersion of the elements of power, which raises possibilities of still different international relationships in the future. Some of the potential changes that could occur in such an environment are: 1) increasing links between prices for raw materials and industrial goods from the West; 2) a restriction of access to resources by the controlling nation if fears of a particular resource's rate of depletion (Kuwait, Venezuela) or concerns about the social consequences of a rapid oil boom (Norway) lead to a reduced level of production; 3) the potential rise of nationalism on a continental or regional basis as the key determinant in new energy-sharing or barter agreements; 4) emerging levels of trade between countries such as Brazil and Nigeria which could presage a change in the flow of exports with potentially damaging consequences for their historic markets, e.g., Nigerian oil to the United States; 5) the potential for conflict among resource-rich and resource-poor countries such as LDCs versus OPEC, South Africa versus the resource-poor countries of black Africa, Indonesia versus Australia, Brazil versus Argentina; 6) conflicts between producers of raw materials, including oil, e.g., Saudi Arabia-Iraq-Iran; 7) the geopolitical implications of a nuclear suppliers' (technology) or a uranium ore producers' cartel; 8) the power that a diplomatically isolated state such as South Africa might be able to employ through effective use of its vast uranium resources; 9) the effect that the rise of economic and political nationalism in the third world could have on the nuclear proliferation issue, which would have its own repercussions on the development of nuclear power for civil purposes.

Could it be that the German-Brazilian nuclear deal of 1975, whereby Brazil insisted on receiving all elements of the nuclear fuel cycle, giving it the capacity to build a nuclear weapon, is a precursor of things to come? Will a uranium ore producers' or technology suppliers' cartel emerge that would deny access to nonnuclear states unless they agree either to pay stipulated prices for uranium and/or uranium technology or agree to stringent weapons safeguards, thus infringing on their national sovereignty or perpetuating a circumstance of nuclear "haves" versus "have nots"?

The above are illustrative of the dynamic geopolitical changes that could confront the world system. In this interplay, security factors will perform a vital role as some new oil-producing states vie for hegemonic status in their respective regional arenas. In this environment, the terms under which those who control resources make them available to those who depend on them will reflect the changing international environment, national foreign policy objectives, and the new loci of economic and political power.

Although a few countries or regions may temporarily reduce their dependence on OPEC oil, this increase in energy supply may not be sustained; it will almost certainly not alter permanently the general dependence upon Middle East supply through the rest of this century. Indeed, it is ironic that a temporary achievement in reducing import dependence on OPEC oil may well lead to a state of complacency which inhibits the development of the very alternative energy sources that will be essential beyond the turn of the century.[11] Specific reference to the energy needs of Europe, Japan, and the United States may make this point clearer.

WESTERN EUROPE

A recent study by British Petroleum for an energy forum held by a European-Atlantic group is illustrative. According to this study, while Western Europe's indigenous oil production may rise to 7 MMB/D by the mid-to-late 1980s, cutting its dependence on imported oil from 97 percent in 1975 to 60 percent, production will begin to fall shortly thereafter, leading to a resumed import level of over 90 percent by the end of the century. By that time, Europe could be seeking as much as 20 MMB/D or 70 percent of OPEC's expected productive capacity, whereas last year it used only 45 percent. By 1990-1995, as European oil-import dependence begins to rise, world oil production from conventional sources could begin to decline. Furthermore, in this period a major portion of OPEC production will remain with Saudi Arabia. It then becomes apparent that current American policy centering on the development of a special relationship with Saudi Arabia could come to be regarded as potentially inimical to European and Japanese needs and interests.

In this regard, the scope and character of the United States-Saudi special relationship takes on added importance in that the relationship could be the pivotal link both guaranteeing Saudi security and encouraging the Saudis to continue to produce at or near full capacity. To the extent that the United States' security umbrella could result in a Saudi maximum production rate, then Western European and Japanese interests would be served as long as a large portion of this oil were made available to their markets (as is presently, and historically, the case).

Nevertheless, given the quantitative difference between the energy import dependence of the United States and Western Europe and Japan, its major allies would be uneasy if the United States used its special relationship with Saudi Arabia to secure first its own access to vital oil reserves while seeming to ignore the needs of Europe and Japan. Sensitive issues are involved, including the intentions of the Saudi government, the interests of the American oil companies which still dispose of the great bulk of Saudi oil, under Saudi direction, and the interests of the United States government itself.

Europe faces a series of problems in ensuring its access to energy supplies: the volatile nature of the Middle East; growing uncertainty over the future role of nuclear power; policy preferences and political developments in Norway and Scotland which could affect development of North Sea oil reserves; a declining coal base; and increasing dependence on imports of Soviet gas and Middle Eastern oil.

The unchanging degree of Western Europe's energy dependence is deeply disturbing, particularly in view of the OECD's estimate that, at current growth rates of electrical consumption, Europe will have to add the equivalent of 800 nuclear plants by the end of the century. For every 100 nuclear reactors that are not built, Europe's oil imports will increase by the equivalent of Kuwait's entire annual oil output (some 730 million barrels). If Europe does not develop nuclear power on a massive and timely scale, it will need to import up to 500 million tons of coal a year by the late 1990s or nearly five times current world coal exports. It is estimated that at best, solar, wind, wave, and tidal power will not be able to supply more than 6 to 8 percent of Europe's energy needs by 2000.[12]

The geopolitical consequences of these levels of import dependence are staggering. Clearly, the United States will not be able or willing to export coal to Europe in anywhere near the quantities necessitated by Europe's burgeoning expansion of electric power. Given this, and the huge costs involved in the development of alternative energy sources, it might be to Europe's advantage to expand its bilateral initiatives with the Arab oil producers rather than to divert scarce resources into high-priced energy alternatives. If such arrangements could be negotiated, OPEC could exert a high price in terms of guaranteed arms sales, joint technological development, and joint patent access. But in return, Europe might hope to gain some supply stability and price predictability as a means of providing an easier transition to a new energy resource base. Indeed, if Europe combined such OPEC-directed initiatives within the context of an expanded system of new international economic relationships (China, the less-developed countries) a new middle-power economic bloc might develop to challenge the economic power of the United States and the Soviet Union. (It is worth noting that Peking has actively supported such a development.)[13]

Although it is difficult to perceive such a scenario developing, given Europe's need for the American security umbrella vis-à-vis the Soviet Union, there is no reason for the United States to accept as a necessity the compatibility of Europe's long-term energy interests with its own, or that Europe and the United States need employ constructive and mutually agreeable tactics to assure supply.

JAPAN

The single most important variable in assessing energy projections for Japan is the rate of its economic growth. Because Japan has almost no indigenous sources of energy, the only way it can help alleviate its overriding dependence on imported oil is to develop nuclear power as quickly as possible. LNG imports and increased coal imports can only marginally affect Japan's oil dependence for at least several decades. Because Japan is completely dependent on foreign sources of uranium, however, it cannot end its dependence on foreign energy supply (given current reactor designs) but can only diversify the level of its dependence on Persian Gulf crude-oil reserves.

Japan, however, has encountered several obstacles in its attempts to convert to nuclear power. On the domestic political scene, the conversion to nuclear power has generated opposition from environmental groups who are concerned not only about the dangers of nuclear waste and proliferation but also about the danger posed to nuclear-plant facilities by earthquakes. Because of the limited fresh-water supply in Japan, concern has been expressed over the thermal effects that nuclear-power development could exert on Japan's water supply. The problem is exacerbated by the high humidity in Japan, which is said largely to preclude the use of cooling towers to counter the effects of the thermal pollution generated by nuclear reactors.[14]

As a result of these concerns, plus the rapid increase in capital costs which affect nuclear projects everywhere, a number of nuclear plants have been delayed, forcing Japan to confront the choice of whether to turn to oil-fired stations or to construct coal plants to generate its electricity needs. While oil-fired plants are favored in a pollution-conscious Japan, the oil has to be imported, thus intensifying the level of Japan's oil-import dependence. And, although Japan has some domestic coal reserves, it would still have to import sizable quantities to meet a nuclear shortfall by coal. Yet the choice of coal over oil would help somewhat to diversify the geographic concentration of Japan's energy-import dependence on the Gulf.

On the international scene, Japan's drive to expand its nuclear sector via the breeder reactor and reprocessing of fuel has encountered strong resistance from the Carter administration, which fears the impact that the commercial development of the breeder will have on nuclear weapons proliferation through the introduction of the plutonium fuel cycle. Because Japan is presently totally dependent on shipments of enriched uranium fuel from the United States for its conventional light-water reactors, the slowdown of those shipments as a form of diplomatic pressure on the breeder issue generated alarm in Japan. The Japanese are particularly perplexed, since, under the Ford administration, they were encouraged to develop the breeder to reduce the level of their oil dependence on the Gulf.

Given these uncertain energy options, Japan could by 1990-1995 face an energy-supply crisis of staggering proportions. Even assuming that Japan might, through effective diplomacy, obtain most of the Far East's oil production, by the early 1990s Japan might find that China and the Soviet Union have entered the market as competitors. Moreover, by 1990 Indonesian and Malaysian oil exports may have ceased, as rising domestic consumption and declining oil reserves remove them from the world market. In such a situation, Japan could face the prospect of growing Chinese and Soviet competition over access to oil in the Middle Eastern market.

If such a situation occurred against the backdrop of a Japanese perception that the United States had access to Saudi reserves for its own use and was channeling them to meet its own rising import needs, Japan might be forced to seek out new special relationships with potentially energy-resource rich nations such as the Soviet Union or Australia. A decision to do so would have strategic implications far transcending present questions of access to energy. The massive participation by Japan in the joint exploitation and development of Siberian energy and mineral resources could imply a major shift in the world balance of power.

Although there are political obstacles (northern islands issue, Japan's prowestern orientation, the nature of Japanese commercial-export markets, the effect of such a development on Tokyo-Peking relations) that would hinder such an alliance, there is a serious shortcoming in western strategic doctrine which assumes that future relations between Japan, Western Europe, and the United States will be continuations of post-1945 relations.

When one considers the tremendous changes that have occurred in the geostrategic environment in the last two decades (Sino-Soviet rift, the expansion of Soviet-European trade to the point that the Soviet Union now provides 40 percent of Western Europe's imports of enriched uranium, the Sino-American efforts at rapprochement), there is little reason to believe that equally momentous changes will not occur during the remainder of the century, and the insatiable need for energy could be a primary cause.

THE UNITED STATES

I am concentrating in this review on the energy needs of the industrial nations, not on the requirements of the developing world. I do so because the magnitude of the requirements of the developed nations dwarf those of the LDCs and by themselves create that ever-increasing pressure upon supply which contains the ingredients for the most corrosive kinds of competition.

The United States is and will remain throughout this century the energy-producing and energy-consuming colossus. It is likely also to continue to be by far the largest single importer of oil while it is almost alone among the industrial states in the wealth of energy options of which advantage has yet to be taken. It is a prescient observation, commonly shared by energy analysts abroad, that the scale of American energy demand, measured in terms of the level of imported oil, may be one of the prime determinants in shaping the coming competition for oil placed in world trade. Presently, the United States is approaching a point at which one-half of its oil consumption will be met from foreign sources. Unless the most stringent and disciplined energy policies are implemented—and soon—that import level could exceed 12 MMB/D by 1985, approaching total European consumption in 1977. The additional 4 MMB/D—added to Europe's and Japan's needs—could be the cumulative requirement for oil that warns of grave trouble. By 1985 it is possible that the present surplus will have disappeared and we shall be on the knife-edge of a supply/demand balance not determined by market forces alone but by physical limitations to supply and politically determined limits to production.

The following tables are instructive both in emphasizing the "energy weight" of the United States and in stressing the period in which the time of unprecedented competition approaches:

TABLE 1
Energy Consumption
(MMB/DOE)

	1976	1980	1985
Western Europe	24	27	32
Japan	7	9	12
United States	37	42	49

TABLE 2
Estimated Oil Demand and Supply
(MMB/D)

	1977	1980	1985
Free world oil demand	50	55	70
Western Europe	(14.0)	(14.2)	(17.0)
Japan	(5.3)	(6.4)	(8.4)
United States	(18.0)	(20.0)	(24.0)
Non-OPEC supply	18.5	22.0	21.4
Required OPEC production	31.6	34.0	48.0
OPEC production capacity Projection			
(less Saudi Arabia)	27.0	28.0	29.0
Saudi Arabia	10.0	15.0	20.0

The inability to change these estimates by dint of energy investments made now results from the long lead times required to bring new efforts into use. A new and major oil field may require five to eight years to develop; a nuclear generating plant may require ten years. (And the discovery rate of oil reserves outside the Middle East has been insufficient to begin reducing dependence upon the Middle East.)

In all these considerations, the point remains that the United States possesses abundant coal resources, a probably substantial natural-gas resource not yet tapped, and substantial oil reserves yet to be discovered; moreover, it possesses one of the largest uranium-ore reserves in the world. Not one of these resources can result in a significant lessening of its demand for oil in world trade—which means increasingly from the Middle East—within the next decade at least. But the potential exists for doing so if sufficient and sustained attention is given to conservation, energy research and development, and to the development of known and presumed domestic energy resources. The launching of such an energy program would in itself be a compelling signal to oil-exporting countries. The United States would be moving to regain an energy self-sufficiency necessary to enhance its own security and to regain a flexibility in international affairs which is denied when its dependence upon foreign energy supply is so marked.

The United States does not live in an energy world of its own; its security depends in part upon the energy situations and prospects of its allies who lack the energy options available to it. Two instances in which insensitivity or plain ignorance on the part of government complicate greatly its alliance relationships are the past denial of Alaskan crude to Japan and the United States' position on further development of the breeder reactor.

In the first instance, Japan has sought to multiply its sources of oil to minimize the risk it runs of a political action in the Middle East threatening its continuity of supply. Japan sought temporary relief from a portion of its dependence by receiving that portion of Alaskan crude considered surplus to American West Coast requirements. There were complicated reasons why the United States was unlikely to agree to such a disposition, reasons that the Japanese would not readily understand. But when the president publicly foreclosed the Japanese option, not in terms of Alaskan oil going only to the United States but with specific reference to Japan, the reaction could only have been expected, and so easily avoided. The Japanese believed they had been singled out and rebuffed in a matter—oil supply diversification—which was surely of common interest to both countries. Yet in the first postembargo example of an opportunity for the United States to demonstrate its awareness of its alliances' dependence upon assured supply, it performed as other nations suspect it always will in matters of this sort.

Even more consequential is a failure in United States energy diplomacy that has come with the administration's nuclear-fuels policy. Europe, Japan, and a number of LDCs see no other alternative to continuing and increasing energy dependence upon imported-oil supplies largely from the Middle East except by taking the nuclear route. Presently nuclear-power sources outside the communist world depend upon American sources of enriched uranium for nuclear fuel. Dependence on any nation for such critical supply is to be avoided, if possible. Such avoidance could come from drawing on uranium deposits (and enrichment services) located other than in the United States, alternatives that are not presently available.

The very record of the United States in this matter heightens uneasiness. It would prefer obviously to be regarded as a reliable

supplier of enriched fuel, yet its uranium ore policies, its opening
and closing of order books, and its reconsideration of some previous supply undertakings were among the factors that led a number of European countries and Japan to embark upon the breeder
technology which offers the prospect of virtually freeing them from
excessive reliance upon imported supply. Yet this goal did not appear so consequential to the United States when it sought to limit
further development of the breeder technology. While this question
is discussed later, the point emphasized is that the United States,
itself possessed of unrealized energy options to limit and even reduce its dependence upon imported oil, has demonstrated a highly
unfortunate insensitivity to the concerns of its major allies. As the
character of the anticipated competition for oil will be shaped by
policies and actions taken now, these early signals from the United
States, which ought to be in the forefront of energy diplomacy—
and has claimed to be—are disturbing.

2

The Concession System
& the Oil Host Governments

OIL PRICES and production levels—the two basic conditions affecting supply—are now less dependent on economic factors than are other commodities and have become increasingly politicized. Why this has happened requires a brief review of the concession system, how it evolved, and how oil price and production decisions were made during the early years of the international oil industry when much of the oil in world trade was produced from colonies or territories controlled or greatly influenced by Britain, France, and the United States.

We are reminded daily how the legacy of this experience still influences current oil discussions and affects our access to vital petroleum supplies now and in the future. The examples of Mexico, Iran, and Venezuela are illustrative because each in its own way and time contributed to the gradual shift in the control of oil from the international oil companies, supported by their own governments, to the governments of the producing countries.

By 1900, the United States and Russia dominated world supply, producing more than 90 percent of the world's oil. Production in the United States vastly exceeded consumption, with the result that many American firms began to market petroleum products abroad. Indeed, until World War I, American companies supplied about one-quarter of total foreign consumption.

During the second decade of the twentieth century, the structure of the international oil industry began to change as American consumption quickly increased and American firms became increasingly dependent on foreign sources of crude to supply their international markets. Still, the transition from domestic self-sufficiency to ever-increasing dependence on foreign crude-oil sources was slow; on the eve of World War I, United States companies possessed oil-producing properties only in Mexico and Rumania, and foreign production by American companies accounted for only 15 percent of total crude-oil output outside the continental United States.

THE RACE FOR OIL

The discovery of the giant oil reserves of the Middle East began in the early years of this century, precipitating a continuing rivalry primarily among British, French, and American companies, backed by their governments, to assure each acquired a significant share in the region's oil treasure. In this rivalry, Great Britain had the initial advantage, seized it, and remained the principal actor in oil until after World War II. Not only had the British long recognized the Middle East as the strategic gateway and communications route to India and the Far East, they had also realized the military and economic significance of petroleum as a new factor in the geostrategic balance.

A major precipitating factor behind the British government's interest in oil was the decision by Winston Churchill, first lord of the admiralty, to convert the British navy from coal to oil immediately prior to World War I. Out of a concern for British control over the sizable reserves of Persia (Iran), London not only moved to acquire a major interest in the Anglo-Persian Oil Company but also sought to influence oil supply in the Middle East generally and extended its interest to involve the oil-producing regions of the Far East.

After World War I, the British Empire emerged from the war not only intact but enhanced in scope, particularly in the Middle East where the breakup of the Ottoman Empire and the end of German interests in the area led to a significant extension of British influence and power, largely through the lands mandated to British control by the League of Nations. British influence over oil was pre-eminent.

In the aftermath of World War I, increased demand and competition within the petroleum industry led to a rivalry among the major industrial countries and their respective companies which had, as one of its results, a renewed American interest in greater access to the oil reserves of the Middle East. Doubts had begun to emerge in the United States concerning the adequacy of the nation's oil reserves. These fears were accentuated when neighboring Mexico, the only major foreign oil-producing country in which there were substantial American oil interests, adopted policies that threatened American oil investments, raising the specter of heightened United States dependence on its own resources.

Concern over the inadequacy of American access to vital foreign petroleum reserves and growing alarm at the predominant role exercised by the British in tying up the world's crude-oil reserves, led in the 1920s to a series of initiatives to ensure that the United States would have access to the sizable oil reserves of the Gulf region of the Middle East. The most significant action the United States government took was to gain access for American companies in the development of the oil reserves of Iraq. Although the measures taken to effect this entry were complex and occurred over many years, by 1928 a consortium of American companies did secure a 23.75 percent share in the Turkish Petroleum Company (TPC) which controlled Iraqi oil resources.

Under the terms of the agreement, however, the American companies were forced to agree not to seek further concessions within an area demarcated by the famous "Red Line" except through the TPC. This area included the former Ottoman territories beyond the boundaries of Iraq. Of the original consortium of American companies, only five were represented in the final settlement with TPC (later changed to the Iraq Petroleum Company—IPC). These were Atlantic, Gulf, Standard Oil Company of New Jersey (now Exxon), Standard Oil of Indiana, and (what is now) Mobil. Despite the settlement, however, oil was not developed in any quantity until 1934, by which time Standard Oil of Indiana,

Gulf, and Atlantic had sold their interests to Exxon and Mobil, who became sole owners of the American 23.75 percent interest.

By the late 1920s, the exclusionary policies of the British in the Middle East and the Dutch in the East Indies had led the American companies into the Central and South American markets. Subsequently, growing fears of a world petroleum shortage had been allayed by a combination of factors including sizable discoveries in Venezuela, the renewal of exports from the Soviet Union, and an economic recession in Europe, all of which led to a weakening of world petroleum demand. These events threatened a disorderly market in world oil and three members of the IPC consortium (Shell, British Petroleum, and Exxon) concluded the 1928 Achnacarry Agreement under which they agreed to take such actions outside the United States which would limit the competition among them.[1]

By the early 1930s further significant changes had occurred in the international oil market that were to have profound implications on the geopolitics of energy. While foreign crude-oil production doubled between 1920 and 1930, the foreign production of American firms rose by only 15 percent because American companies had virtually no interests in the new crude supplies of the eastern hemisphere. Likewise, while foreign consumption remained relatively stable, American consumption more than doubled during this period, and foreign production supplied about two-thirds of world (outside the United States) consumption. Finally, by 1929 Venezuela had become the largest exporter of oil in the world.

THE MIDDLE EAST AFTER 1930

By the early 1930s, world oil-producing capacity had swung full circle from the fears of depletion in the early 1920s. By 1931, with the East Texas fields in production, domestic crude-oil prices plummeted. By 1933 the world glut of petroleum, arising from depression-reduced industrial activity, had caused United States exports of refined products to drop by 50 percent and the American export share of foreign consumption to fall from 30 percent to some 17 percent. Indeed, so poor were the prospects for international sales that a number of American companies withdrew from the international market.[2]

For the American companies that stayed in or entered the market, however, the 1930s were critical years in establishing their access to vital crude-oil reserves. In the early 1930s, Standard Oil of California (Socal), through a series of fortuitous events and shrewd bargaining, breached the British-dominated Red Line and gained a vital concession in Bahrain. The discovery of oil there in 1932 generated renewed interest in Saudi Arabia and Kuwait, where, by a succession of political and economic maneuvers, Socal, Texaco (in Saudi Arabia), and Gulf (in Kuwait) established themselves as successful competitors to the British.

THE CONCESSIONS

Access to oil was expressed through a general system of concession arrangements, backed by the home governments of the companies, a support that was essential to the creation and durability of the oil system. Although each concessionary arrangement varied somewhat from the others, in general a typical concessionary arrangement often included payment by the companies to the producing government which was repaid to the companies when production commenced; a nominal rent for the designated area; and a fixed royalty payment on the oil produced. The companies were usually exempt from other taxes and import duties. In return, the companies sometimes agreed to an annual minimum payment to the government. The concession arrangements were of long duration, sometimes for sixty to seventy-five years. The concessions were often quite large, and the rate of exploration and development was almost wholly at the companies' discretion.

Although the terms of the concession system were to change dramatically after World War II, as the producing governments shortened lease terms, obtained their own shares of production, and received greater revenue, the companies still maintained effective control in all respects. In general, payments to producing governments averaged between 10 and 30 percent of the value of a barrel of oil produced in the United States. Although prices and profits continually rose abroad as demand sent oil prices spiraling, the returns to producing governments remained fixed at a very low rate—in some instances only four gold shillings per ton.

Prior to the war the major portion of international oil trade centered primarily on the movement of petroleum products (crude oil forming only one-third of the total volume); after 1945 the market changed fundamentally. Indeed, while between 1950 and 1972 the volume of international trade in crude oil and petroleum products rose from 2.9 to 30 MMB/D, the component of crude oil in this trade rose from 33 percent in 1945 to 50 percent in 1950 and 83 percent by 1972.[3]

In addition to the unprecedented rise in the total volume of crude oil traded, the direction of international petroleum trade also changed dramatically. In the immediate postwar era, the primary flows of petroleum were from South America into the United States and Europe and from the Middle East into Europe and the Far East. By the early 1960s, and on into the 1970s, important shifts occurred. By the early 1960s not only had a significant level of petroleum trade developed between the Middle East and Canada and the United States but North Africa as well began moving sizable quantities of crude. Middle Eastern crude exports to Southeast Asia and Japan also grew rapidly.

At the time of the 1973-1974 Arab oil embargo, the Gulf had gained in importance from a relatively minor role in international oil trade to a position of overwhelming predominance. This rapid rise in production levels vastly enhanced the geopolitical importance of this part of the world and also demonstrated that the fledgling nation-states of the area, if organized, could yield vast political and economic influence.

THE BASIS FOR CONFRONTATION

Prior to World War II, revenues from oil for the producing governments depended upon fixed fees per ton or barrel of oil. Under this arrangement, producing governments had very little interest in either real prices or costs. As the concession system began to change, however, influenced by the winds of political change and the loss of empires, the international oil industry was also transformed. The major factor in this transformation was the achievement of political independence by the former colonies, protectorates, and mandated territories—and the realization of these independent producing nations that they could assert sovereignty over the disposition of their natural resources.

The oil companies had controlled the oil in each of the producing states. They were able to exert great influence over the political and fiscal policies of the oil-producing states and they were influential in shaping the direction and scale of developmental plans and programs. Likewise, because of the companies' close association with the colonial governments in the early days of the concession system, the companies were perceived by third-world nationalists as commercial extensions of the authority of the companies' home governments. As a result, the concession system came under increasing attack not only because the terms were lopsided in favor of the companies but also because the concession system symbolized the practical continuation of colonialism even after political freedom had been achieved. Thus, even after most of the oil-producing states thought they had achieved full sovereignty, the perpetuation of the concession system remained a constant reminder of an earlier, dependent colonial-style relationship.

Because of this legacy, the newly emancipated oil producers soon began to press for a greater share of the benefits from the exploitation of their oil. Although the international oil companies remained the most important economic forces in all the oil-producing states, the "majors" soon found themselves assailed not only by the governments of the oil-producing states but also by the smaller, new companies, the "independents," who agreed to give the oil-producing states better terms in exchange for access to vital crude supplies. Moreover, in every producing state, national oil companies, created by governments, began to replace the majors, first in domestic supply and then in exploration (often in joint ventures with the independents).

Under the impact of a growing government role, there developed, in addition to royalty payments, a system where payments were derived from a percentage of the profits; as this occurred, the oil-producing governments had a vital interest not only in the volume of oil produced but also in the cost of production and its sale prices to affiliates of the majors and to third parties.

With the change to a system whereby the producer governments' take was to a large extent determined by the profits of the concession holders, a price yardstick was needed against which different kinds of sales could be measured. A "posted price" in the Gulf was devised, the origin of which lay in a public price mechanism developed by the oil companies on which prices for oil sold anywhere were based on American prices. Increasingly, however, it was to be a matter of dispute as to whether changes in the posted price could be made unilaterally by the companies or whether approval by producer governments was necessary.

In the context of these complicated issues, it became commonplace to attack the profit realized by the oil companies, overlooking the beneficial aspects of the companies' presence in the oil-producing states. As providers of new technology, capital, and marketing facilities, the companies often served as agents of economic and social change. By establishing schools to train local personnel, the companies provided new educational opportunities. This is not to argue that the companies' activities were altruistic. The disparities of income and opportunity that continue to plague oil-producing states, however, were far more the result of inadequate education and institutions and the rigid hierarchical social structures than a product of oil company influence. Nevertheless, when one compares the revenue accruing to government with the large share retained by the companies, adjustment was long overdue, and when it did come, it was mostly brought about by the producing governments and forced upon the enterprises.

Still more fundamental changes were to come. One of the most consequential occurred in Indonesia under the direction of General Ibnu, who produced a major and farsighted set of policies. It was General Ibnu's conviction that the political connotations of a concession system whose origins lay in the Dutch colonial period could no longer be accepted. He compelled the majors to accept a variety of other arrangements generally embraced within the concepts of "service contracts" or "production-sharing" agreements. The common denominator of these was ownership of the oil by Indonesia, with the contracts between oil companies and PERTAMINA, the Indonesian government oil monopoly—a far different relationship from the one embedded in the concession system.

In earlier periods, these major challenges to the international oil system might have been met by a military response or "gunboat diplomacy." That they were not—although the Iranian situation may have come close to it—was due to the postwar decline of European power, combined with European and American reluctance to consider such measures in that political environment (including some concern about a Soviet response) and the basic common sense of the companies. No event made these factors more evident than the defeat of the punitive Suez Expedition of 1956 and the American refusal to stand aside when the British and French attacked Egyptian forces. From that time on, if anything more was needed, the oil producers had every reason to believe that they could press their demands upon the companies without fear of a military reaction to preserve a western commercial interest.

MEXICO

Mexico has a special legacy in the history of oil producer-consumer confrontation, since, unlike the rest of the producing world, this confrontation occurred before World War II. Although oil was first produced in Mexico around 1900, it was the oil requirements of World War I, combined with improvements in technology, that brought about a serious large-scale interest in Mexico. Indeed, Mexican production rose from 72,000 B/D in 1914 to 560,000 B/D in 1920. Most of this production was the result of British and American initiatives and was considered to be owned by their oil companies.

The period from 1910 to 1920 was also an era of great unrest as the Mexican people, frustrated by a political, social, and economic structure that had changed little in 350 years, moved to assert their rights to a better life. In 1910 a revolution against Porfirio Diaz, dictator for the previous thirty years, unleashed ten years of civil war. Nonetheless, on May 1, 1917, a new constitution was promulgated which embodied into law the sweeping reforms that had been the result of social upheaval.

The constitution included two provisions that directly affected the international oil industry and were to become the subject of controversy for the next twenty years. Under Article 27, the direct ownership of resources, including petroleum and all solid, liquid, and gaseous hydrocarbons, was vested in the state. In addition, the constitution defined the right of the state to expropriate such properties, with compensation, if such action were deemed in the national interest.

The new constitution generated alarm in both private and governmental circles in the United States because the Mexican action posed a direct challenge to continued American access to Mexican oil reserves. Controversy arose because the interpretations of the Mexican and American governments toward these constitutional provisions differed widely. While the United States' interpretation was that companies operating prior to May 1, 1917, continued to own the oil deposits, Mexico dissented, arguing that the companies had vested rights to explore and produce but not to own the oil reservoirs.

The problem was exacerbated by the legacy of American intervention in Mexico, especially United States support for the ousted Diaz dictatorship, which made the oil dispute a test case for the new Mexican nationalism. The conflict was further fueled when the United States government actively entered into the dispute and exerted strong pressures on various Mexican administrations that were often too weak to resist. (In this regard, it is interesting to compare American intervention in Mexico with the growing demands in the 1970s for a more active United States government involvement in all aspects of international petroleum negotiations. It is worth recalling that active government intervention in the negotiations in Mexico only served further to polarize the already tense relationship between the oil companies and their host government. While the companies were far from being ready to accommodate to change, their attitude was stiffened to a rigid position by the government support. In the end, the companies lost everything.)

Throughout the late 1920s and 1930s, Mexican-American relations remained bad as growing concern over rising Mexican taxes and the ultimate security of ownership led to a plummeting in oil production levels from 560,000 B/D in 1920 to about 85,000 B/D in 1930 and about 105,000 B/D in 1938.[4] The volatile Mexican political situation, combined with the discovery of sizable new oil fields in the United States, Venezuela, and Colombia, led by the late 1930s to a further deterioration in Mexico's oil prospects. Matters finally reached a boiling point when the Mexican government supported the petroleum workers in a labor dispute with the foreign oil companies. Beset by the greater bargaining power of the companies in an era of world crude-oil surplus, which allowed the companies to adopt a hard line due to their ability to switch to other crude-oil markets, the Mexican government, irritated and threatened by the active United States government role on behalf of the companies, announced on March 18, 1938, the complete expropriation of all foreign oil holdings.

Since the full nationalization of foreign oil holdings in 1938, the Mexican oil industry has gone through a difficult period as the removal of Mexican oil from the integrated networks of the international oil companies drastically cut sales and development efforts. During and after World War II, production recovered and increased. Nevertheless, ever-rising domestic demand outstripped production, and Mexico became a net oil importer in the early 1970s. Relations with the international oil companies were confined largely to purchases of oil and equipment.

In 1974 new discoveries raised hopes that Mexico would once again become a net oil exporter. Since 1974, estimates of these reserves have been increased to the point that current Mexican reserves, as of late 1977, are estimated by some to be as high as 60 billion barrels, or possibly six times greater than that of the North Slope fields of Alaska, and probably greater by far (on the basis of conservative estimates) than the North Sea reserves as presently calculated.

Even though there is still great uncertainty as to the size of Mexico's deposits, the possible existence of such sizable oil reserves in Mexico has generated great American interest not only because the United States is interested in oil discoveries wherever they occur but also because it is argued that their proximity to American markets makes them more reliable from a national security viewpoint. In addition, some strategic analysts believe that any potential diversification of American imports away from the Middle East should be pursued vigorously, since such diversification lowers the level of competition with its allies for Gulf crude and restores a flexibility to its actions overseas. Although such supply diversification would be fortuitous for the overall American energy situation, particularly because of the low sulfur composition of most Mexican crude, the history of United States-Mexican petroleum and governmental relations (not to speak of the legacy of mistrust of the companies) gives no reason to be confident about the prospects for successful United States entry into the Mexican market.

There are several consequences of this legacy which are certain to affect access to Mexican oil, some of which relate not only to past experiences but to present Mexican needs. Combined, these factors will shape the future pattern of Mexican oil-export sales. On the one hand, Mexico's continuing economic crisis is clearly pressuring the regime of President Lopez Portillo drastically to speed up the exploitation of the country's oil reserves so that added revenue, generated by increased export sales, will be available to the Mexican government's developmental programs and to meet Mexico's staggering loan obligations. On the other hand, in order to achieve this rapid development, the Mexican government will be dependent on foreign technology, sources of financing, equipment, and probably management, which raise political problems with those groups whose political attitudes are governed in part by the legacy of the 1938 nationalization. These political groups do not want the international oil companies invited back into Mexico, except possibly as contractors with no equity interest in Mexican oil. They see no reason why American firms should lift Mexican crude and acquire the value added through refining but believe that Mexico should limit its crude exports so that it can market increasingly the more valuable refined products.

Although the fact that Mexico has not yet joined OPEC is often cited as an indication that the Portillo regime favors an expansion in United States–Mexican petroleum relations, in reality, the geo-political implications of Mexican reentry into the world oil market are far more complex. A Mexican decision to join OPEC would ef-fectively exclude Mexico from the benefits (trade preferences) of the 1974 United States Trade Expansion Act and would inevitably ex-acerbate tensions with the United States. Mexico can continue to benefit from high oil prices without accruing the headaches of for-mal OPEC membership. This fact has not been lost on Saudi Arabia, Iran, or Venezuela who are uneasy over how the advent of Mexican oil in sizable quantities into the international marketplace could affect their market shares in the United States. If Mexico's production does rise to the most optimistic predicted levels, Mexico will undoubtedly find itself under greater pressure from OPEC to join the organization because of the prospect that most of Mexico's oil might flow to the United States.

Such a large flow (conceivably, by the late 1980s, 2 to 3 MMB/D or some 20-25 percent of United States daily import needs) of Mexican oil into the American market would have obvious ramifications, since the United States would have partially diverted the source of its imports away from the Middle East, thus reducing somewhat the potential for conflict with its major European allies and the Japanese over Middle East supply. But if it is assumed that the magnitude of the proved and probable reserves is in the 60 bil-lion barrel range, Mexico would still need to invest some $15 billion over the next six years to develop them. If, as Mexico argues, they would then, by 1980, be producing only 2.2 MMB/D, this would allow an export potential of some 1.1 MMB/D which would have little impact on the world petroleum market. Crucial to these fore-casts are the sufficiency of reserves, the availability of technology, whether the requisite managerial skills need to be imported, and finally, whether Mexico can generate additional loans or invest-ments of the necessary magnitude. Despite these problems Mexico will probably be able to attract United States and other foreign in-vestment, given the proximity of Mexico and the probable quantity and quality of its crude. Even if all these factors become positive, however, they do not imply that Mexico need necessarily favor the

American market. Access to Mexico's oil—as in the case of many other producers—will not be determined by market forces alone.

VENEZUELA

As petroleum demand increased during and after World War I, those major oil-consuming countries possessed of the means to do so sought access to overseas petroleum.[5] While the competition eventually centered on the Middle East, Venezuela was an early and important target. In the aftermath of World War I, Exxon and Gulf entered the competition, largely through other companies already present; and Shell was also active. From a position of a comparatively unimportant producer of oil in 1922, Venezuela became in less than a decade the world's largest exporter of petroleum.

Throughout the long history of foreign exploitation, the ownership of Venezuela's petroleum rested firmly with the Venezuelan government as a consequence of its legal tradition, which held that subsoil resources fall under the permanent ownership of the state. This conviction resulted in the companies' possessing exploitation rights but not ownership of the reserves. Payments to government were for the development of oil resources and their eventual sale in the market. The arguments that developed in Venezuela with the companies, however, and that occurred later in the Middle East, were over the issue of government revenue: was it a fair share? The force of Venezuela's argument was lodged in the inescapable fact that economic development and social progress depended totally on the realization of the greatest possible return from the companies' exploitation of oil. The politics of Venezuelan oil were engaged in the running controversy which lasted nearly a half-century and still affect the country's attitude toward the foreign companies, even after nationalization. It was a long and difficult struggle between government and the oil industry, made more complicated by the growth of lower-cost oil from the Middle East concessions held by the same giant oil companies active in Venezuela.

As was the case in all producing countries, Venezuelans lacked initially the requisite knowledge to oversee or participate in oil operations. Accordingly, the earlier years of oil exploitation found the government increasingly possessed of the need to learn and to obtain far more knowledge about the profitability of the companies and what revenues could be obtained for the country. Without other sources of information, at least in the earlier years, the advice of the oil companies was sought on laws and regulations. Nevertheless, criticism of the companies mounted, muted in some years by a close and, as some Venezuelans perceived it, corrupting relationship between the industry and periodic dictatorial regimes. Still, the general political trend deepened: the Venezuelan government, reflecting the views of political groups, began to apply higher tax rates on company profits, to examine the books of the companies, to oversee the costs of producing Venezuelan oil and the prices charged by the companies, and, belatedly, to understand the implications of the higher-cost Venezuelan crude—not only to the companies but to the Venezuelan economy itself. At the same time, the Venezuelans were aware of the attractiveness of their oil and were knowledgeable of the extraordinary role of Venezuelan profits to the Standard Oil Company (now Exxon). There were years when Venezuela alone accounted for one-fourth to one-third of all profits of Standard Oil of New Jersey in its worldwide operations. The major market became the East Coast of the United States.

The great importance of the American market to Venezuela set the stage for a prolonged search by Venezuela for a special or preferred relationship with the United States, a search that received little support from the Department of State, which was persuaded that encouragement of Caracas by Washington could only complicate the commercial operations of the companies and dilute their paramount role. Even during World War II, when petroleum demand soared and the Venezuelan oil source was of vital national interest, the government in Caracas was not able to alter the kind of relationship it had with the United States. After the war, with even less prospect for a change in the external form, the Venezuelan Congress returned its attention to the tax rates and the issue of whether concession areas were being fully and fairly exploited and whether new areas would be made available to the companies.

Large as the Venezuelan contribution was to the world oil

trade in those earlier years, it was becoming evident that the Middle East was quickly outstripping all other exporting regions. The Venezuelan reserves of conventional crude (excepting the heavy tar belt of the Orinoco Valley) would in the foreseeable future play out—a time that would come even sooner if the Venezuelans overplayed their hand and the international oil companies showed no further interest in exploration and exploitation.

The great watershed in Venezuelan relations with the oil industry came in 1948 with the imposition of the "50-50" profit tax formula. The political return of this victory was incalculable; for the first time Venezuelans could express in (overly) simple terms the equality of the relationship. The balance previously tipped in favor of the companies had been righted. The presumption was that the scale would now increasingly favor Venezuela; time and revenue would now be made up for the prior years.

The role of Perez Alfonso in this and later periods was of greatest importance to Venezuela and, indeed, to all oil-producer governments. Knowledgeable, politically sagacious, persistent, and gifted in the arts of negotiating and leadership, he stands out as perhaps the greatest architect of the international oil system (from the producers' viewpoint). The 50-50 agreement spread from Venezuela and came to affect (*infect* would be the term used by the industry) oil relationships worldwide. Moreover, Venezuela began an active search for ways and means to deepen its knowledge of all aspects of the oil trade. Subsequently it began to apply its growing skills to the marketing of some of its crude (although the companies sought to limit this role by buying back from the government the oil that would otherwise put Caracas into the oil business).

The next major advance in the assertion of Venezuelan control over the disposition of its resources came in the mid-1950s with the successful imposition of retroactive tax payments. This occurred while some producers in the Middle East were extracting advance tax payments; the difference between the two approaches illustrates the heavier clout of Venezuela and the greater artfulness of its tactics. By the end of the decade, the Venezuelan Oil Corporation was established to become the primary instrument of Venezuelan government oil operations.

The world oil market was then in surplus and the international majors felt compelled to reduce their posted price, which they did unilaterally, without the agreement of producer governments. Because oil agreements had long since moved from a flat government take per ton, in which the governments received a flat fee regardless of actual price or profit, Venezuela and several other countries saw, in the companies' action in reducing the posted price, an immediate threat to their revenues. The result—the formation of OPEC—was to change profoundly the relationships among the consuming/importing governments, producing/exporting governments, and the international oil industry. But ten years were to elapse from the founding of OPEC before the tentative groping for producer unity in terms of oil would be transformed into the extraordinary set of negotiations which began in the end of the 1960s and gained great momentum in the first years of the 1970s.

In the decade of gathering producer strength (ignored by the companies and consuming governments alike), Venezuelans led the way by encouraging producers to learn more from one another and to develop the skills essential to the success of the eventual takeover from the companies. Here, again, the figure of Perez Alfonso looms large. In 1972, faced with an inadequate government revenue from oil and a mounting burden of economic development costs, the Venezuelans exerted strong pressure, insisting that the companies maintain an export level regardless of world oil-market conditions. Full nationalization, in January 1976, was only the capping of a process that had no other objective.

Through this period of great tension, the Venezuelans asserted repeatedly that they were a secure source of supply, that they had continued to provide oil to the United States in every supply crisis, including the Arab embargo of 1973-1974 (and had even offered to increase supply in the winter fuel crisis of 1976-1977). Even though it had been in the forefront of those countries urging massive and continuing price increases, there was little reason to doubt the genuineness of Venezuela's interest in deepening and extending its role as a steady and stable supplier to the United States.

These efforts are thought to have been blocked chiefly by the State Department which was still persuaded that while negotiations—a somewhat inaccurate description of the process whereby Caracas imposed new terms on an embattled industry— were under way between the companies and the Venezuelan government, a signal of preferred source or a burgeoning special relationship would cut the ground from under the companies, as indeed it would have. Nevertheless, the argument should still have been made that access to Venezuelan oil, sooner or later, would be determined not by the companies but by Caracas; and also that a hemispheric source would surely have greater security than sources in the Persian Gulf. Moreover, the still-to-be exploited tar sands could conceivably open up a petroleum resource of staggering size—perhaps some 300 billion barrels "recoverable." But the potential of this resource—possibly of such a magnitude as to surpass anything presently known in the Middle East—can probably be realized only by Venezuelans' calling upon the technical and financial resources of the industrial world.

The United States, and its oil industry, has too easily assumed that the successful exploitation by Venezuela of its Orinoco resource would depend upon American know-how and capital and that consequently, the Venezuelans, faced with a depleting conventional oil reserve, would have to offer the United States highly attractive terms. This is possibly the case. But European and/or Japanese interest could also develop in the Orinoco resource which, without United States participation, might eventually obtain access to this immense reserve. With American assistance it might be accomplished earlier. However, and with whichever partners, the Venezuelan undertaking will be accomplished through government involvement and direction—with companies probably holding service contracts—which will mark the culmination of the shift from company control which the Venezuelans forced upon the industry.

IRAN

Equally instructive in charting the course of accelerating change in access to oil is the example of Iran whose resources and strategic location combined to result in a longer history of involvement in oil than that of any other major producer and exporter of oil.

The Middle East was vital to the British as the key communication and logistics link with its imperial interests in India and the Far East—an importance enhanced by the opening of the Suez Canal. This importance became even more pronounced early in the twentieth century, when the British government, recognizing the growing economic and military significance of petroleum, moved to ensure continued access to oil reserves and aided a British commercial company in acquiring and preserving an exclusive exploration concession to most of what was then Persia. To ensure that the exploration of the concession would proceed in accordance with the British government's interests, Parliament moved in 1914 to acquire a substantial interest in the British Petroleum Company (then the Anglo-Persian Oil Company).

Until the early 1950s, British Petroleum (BP) benefited from its near exclusive rights to Iranian oil. Its predominant position was particularly profitable during the 1945-1950 postwar period when, capitalizing on large production increases arising from its large-scale capital investments, BP was able to market ever-greater volumes of oil. During this period, however, BP was confronted by a political crisis of growing proportions. When the war ended in 1945, Iran was in an extremely difficult political bind. In the north the country was occupied by Soviet forces, while in the south the British government moved to reinforce its privileged position. This situation was greatly complicated by the United States, which had propelled itself into Iranian affairs in order to move wartime supplies through Iran to the Soviet Union.

The political milieu was further complicated by the close relationships between BP and the British government; BP was perceived by a growing number of Iranian nationalists as the commercial arm of British imperialism. These views were reinforced by the company's record in employing Iranians, in the prices it charged for petroleum products in the domestic market, and, most importantly perhaps, in the influence it exerted in Iranian fiscal matters.[6] Moreover, a majority of political forces in the Iranian parliament (Majlis) favored the nationalists who desired to rid Iran of all foreign domination. The event that ultimately precipitated the nationalization crisis of 1951 was the new government-BP agreement of

1949 on domestic prices and fiscal terms. This agreement was rejected by the Majlis in December 1950 as being too favorable to British Petroleum.

THE 1951 OIL NATIONALIZATION CRISIS

In 1951 Iran nationalized its oil resources, due in large part to British Petroleum's refusal to accept the 50-50 profit-sharing agreement conceded by the American oil companies to Venezuela and Saudi Arabia. The reaction of the major oil companies to the Iranian nationalization was to implement a boycott of Iranian oil. Prospective commercial clients were warned that if they purchased oil from Iran they would be subject to legal action by BP, since the company claimed it still legally owned the oil. This boycott brought the country to its knees. Oil exports dropped from over $400 million in 1950 to less than $2 million in the two years between July 1951 and August 1953.[7]

The oil companies were able to survive the crisis because of an existing 1.5 MMB/D surplus productive capacity on the world market. In addition, between 1951 and 1955, Saudi, Iraqi, and Kuwaiti production increases, combined with small increases from other Middle Eastern countries, offset the loss of Iranian production.[8]

Although the political circumstances behind the final overthrow of the Mossadeq government remain subject to debate, a slightly more amenable government, headed by the shah and installed with United States support, eased the crisis in 1954. In its aftermath, the role of BP was changed, and a leading Middle East producer survived the experience. Iran went on to become a key leader in the changing world of international economics and the geopolitics of oil.

One result of the crisis was the replacement of the preeminent role of BP by an international oil consortium, of which BP was the most important member. Another consequence was the creation of the first state-owned oil corporation in the Middle East, the first step in a long process of increased government control over natural resources. And finally, the crisis demonstrated that foreign governments would intervene when they believed that their vital interests were affected and would act to support their national commercial entities in the Middle Eastern oil market. Nevertheless, the full test of governments' protecting their interests was never met: military force was not applied. This lack of a traditional reaction to a threat of great consequence to a major imperial interest was an important, perhaps crucial, lesson of the Iranian crisis. Although the United States' role in the crisis was probably directed more to thwart Soviet influence than to support oil interests, there can be little doubt that the American oil industry, and hence United States security interests, were greatly enhanced by the shah's return to power.

The crisis warned that superpower conflict over access to Middle Eastern oil was a distinct possibility and had to be accounted for in any assessment of the balance of power in the Middle East. It also demonstrated that the United States would not accept its exclusion from any area of traditional, albeit waning, British domination. The inclusion of the independents, even as a token percentage of the consortium's interests, helped stimulate growing competition among both the independents and the majors.

With the Iranian situation stable throughout the 1960s, the National Iranian Oil Company (NIOC) sought to encourage interest in the timely exploration, development, and control of its hydrocarbon reserves. As part of this endeavor, NIOC formed partnerships and concluded contractual agreements with foreign firms, usually maintaining an equity interest. In addition, in 1964, NIOC took complete control of the liquid natural gas industry in Iran, while in the international arena it moved to gain refining interests in India, the Republic of South Africa, Europe, and black Africa.

THE 1970 LIBYAN CRISIS

By 1970 ever-escalating levels of petroleum consumption in the industrialized world, combined with the closure of the Tapline and the Suez Canal, led to a world supply/demand/logistics crisis of growing proportions. The supply shortfall helped achieve success in Libya's radical Qaddafi regime's confrontation with the oil companies over the issue of prices, thus affecting the international petroleum system's balance of power between companies and governments. Libya's success demonstrated to other oil producers that they could again risk the imposition of unilateral price changes upon the companies without being challenged by consumer governments, particularly the United States.

The immediate result of Libya's confrontation with the companies was the 1970 OPEC Conference in Caracas where both a resolution demanding a 55-percent tax rate for all member states and the establishment of a pricing committee of Persian Gulf countries (Gulf Committee) were announced. The Gulf Committee almost immediately demanded negotiations in Teheran with the companies in order to discuss increases in the share of oil revenues accruing to the producer governments.

The 1971 Teheran Agreement, which raised the government take of the Gulf producers about $0.30/barrel, unleashed the pattern of OPEC price-leapfrogging, where an agreement with one set of producers immediately led to still higher demands among other producers, even though in theory each set of negotiations was supposed to stabilize prices rather than raise them to ever-higher levels. This pricing pattern continued throughout the early 1970s until it was overtaken by the 1973-1974 Arab-Israeli War and the spectacular fourfold increase in crude prices imposed by the producers.

Prior to the outbreak of Arab-Israeli hostilities in October 1973, Iran had moved to bring about a profound change in its relationship with the consortium. As defined in a new twenty-year agreement, NIOC assumed full ownership of the consortium's oil facilities, operations, and reserves, while the consortium, through its Oil Service Company, continued its technical assistance role.

IRAN: THE POST-EMBARGO PERIOD

The OAPEC embargo of 1973-1974 was a watershed in international oil relations in that it dramatized not only that cheap energy was a thing of the past but also that security of energy supplies would henceforth be as important as price and that the politics of oil could be consequential. Thus the success of the oil-price rise of the OPEC producers warned that the supply of energy raw materials could be affected increasingly by political forces and that western assumptions concerning the nature of the factors that control access to raw materials would have to be revised. Finally, if more generally, it became apparent that new terms and modes for access to raw materials might play an increasingly important role in the economic development of nations and in the shifting balance of world power.

In the aftermath of the embargo, United States-Iranian relations became ever more complex, since the United States now held Iran to be an important actor in the politics of Middle Eastern oil while still a bulwark against Soviet penetration of the area. Defense of the Persian Gulf remains a cardinal tenet of American foreign policy. Even if the United States were able to reduce the degree of its dependence on Gulf sources of crude, the Gulf will remain of critical strategic importance to American allies in Western Europe and Japan as well as to its own larger security interests.

Most of the oil in international trade passes through the Iranian-controlled Straits of Hormuz on its way to markets in Western Europe, Japan, and the United States. Control of the approaches to the straits by a friendly power is deemed of paramount importance, and, as a result, all post-World War II administrations have encouraged the buildup of Iran as a formidable, middle-grade military power. While no one would question the importance of the straits being controlled by a friendly regional power, the United States-Iranian military relationship complicates American relations with Saudi Arabia. The United States must balance the Iranian relationship with its strategic and economic interests in Saudi Arabia, a country whose vital interests, particularly in oil pricing and production decisions, have often been at variance with Iran's.

Saudi-Iranian relations are extremely complex. While both nations share common interests in opposing the establishment of radical regimes in the Gulf and protecting the flow of oil to the West, there are important differences between them. While Iran remains a militant on oil prices, owing to its need for ever-greater revenue to pay for its ambitious economic development and military programs, Saudi Arabia, with its low absorptive capacity and its vast financial reserves, is said to fear that too rapid an escalation of oil prices will lead to economic chaos in the West which will spill over into the Arab world. Moreover, the Saudis, as members of an Arab country, are suspicious of the (non-Arab) Iranians, particularly of their overwhelming military superiority.

The suspicions, ambitions, and widely varying interests of these two key states, plus their respective relationships with the United States, raise so many political issues likely to affect oil (including their record of different stances toward Israel) that, perhaps, only one point needs to be made: in the Gulf, questions of access, which previously involved the relationships between the oil-producing governments and the companies, have now moved to a wider, more difficult arena involving the international interests and actions among key producer governments themselves. With that development, issues of access—the terms, and whether oil will be used as a political weapon—enlarge our concerns.

IRAQ

Although the United States emerged from World War I as the world's largest producer of crude oil, the demand placed on the nation's oil reserves during the war, combined with acute fears at home that its oil reserves were inadequate to meet its future needs, led the United States government to encourage American oil companies to seek access to foreign oil reserves.

The history of oil exploration in Iraq began in the decade before World War I (1904-1914), when Iraq (then known as Mesopotamia) was controlled by the Ottoman Empire. Petroleum rights to the country were vested in the Turkish Petroleum Company (later changed to Iraq Petroleum Company—IPC), a consortium of German and British commercial companies. As a consequence of World War I, the British obtained control over the German share. Subsequently, in the San Remo Agreement of 1920 the British made room for the French government in the exploitation of Iraqi oil.

Since the United States was concerned about its continued access to adequate petroleum reserves, it used the concession turnover to France to enunciate an open-door policy for American firms to gain access to Middle Eastern oil reserves. Although the British and French interests labored to keep the American firms out, the State Department succeeded, and in 1925 the first concessions to the Iraq Petroleum Company were placed under the ownership of BP (23.75 percent), Shell (23.75 percent), Compagnie Française des Petroles (23.75 percent), Exxon (11.85 percent), Mobil (11.85 percent), and Gulbenkian (5 percent).

Exploratory work commenced almost immediately, culminating in the 1927 discovery of the large Kirkuk field. Because of the tremendous logistical difficulties in moving Iraqi oil, however, commercial production was somewhat retarded. By 1934 daily production was only 21,000 B/D, rising to about 80,000 B/D on the eve of World War II.

Another factor affecting Iraqi production was that by the time the IPC concession began to produce (1934), the international petroleum market was glutted. This situation raised difficult problems for some of the IPC shareholders who were faced with the dilemma that, by increasing production of relatively high-priced Iraqi oil, they would add to world overproduction of crude oil, thus further depressing the world price of oil. British Petroleum and Shell, for example, were apprehensive of the effects that rapid development of Iraqi crude would have on their sizable interests in Iran. Exxon was concerned that the addition of large-scale Iraqi production could force down prices in Europe, which were already threatened by the new crude-sharing agreement between Socal and Texaco and which challenged Exxon in its traditional markets. Mobil was probably the only shareholder who wanted more production from the Gulf area; however, because of its small voting share in the IPC consortium, it was unable to exert sufficient influence to increase production.

The net result of these concerns was that Iraqi production was retarded, while production elsewhere in the Gulf increased. Thus, while in 1936 production in Iran was double that in Iraq, by 1948 Iranian production was seven times greater. Similarly, in Saudi Arabia, although production did not commence until 1938, by 1948 Saudi outproduced Iraq by a factor of six.

Whether IPC's motivations were crassly directed by a desire to manipulate the market to the detriment of end-use consumers, as John Blair has suggested,[9] or by the high cost of moving Iraqi oil to market, combined with other valid commercial considerations, the fact remains that, by the late 1940s IPC's reputation for delaying tactics was deeply rooted in Arab opinion throughout the Middle East.

Although World War II seriously disrupted IPC's operations, the end of the war saw little change in IPC's production. With the development of the Kuwaiti and Saudi concessions in the postwar period, the American members of the IPC gained access to large-scale crude-oil reserves that were far more important to them than was Iraq where their ownership interest was limited to only 23.7 percent and where the relative costs of production still made increased production economically unattractive.

Disturbed by growing evidence of IPC's curtailment strategy and the crisis surrounding the Iranian nationalization (May 1951 to October 1954), the Iraqi government on June 29, 1953, concluded with Saudi Arabia an agreement under whose terms Iraq and Saudi Arabia would exchange information relating to petroleum and would hold periodic consultations concerning petroleum policies. The Iraqi-Saudi agreement was important because it signaled the first attempt by producers to harmonize their policies in order to obtain best term clauses from oil concessionaires. Under the principles established by this agreement, the producer governments enunciated the right to renegotiate with their concessionaires for possible revisions in existing agreements if neighboring countries obtained better terms.

Iraq continued nevertheless to be plagued by the economics of its petroleum which placed it at a competitive disadvantage with other Gulf crudes. Indeed, even after the closing of the Suez Canal in 1956, northern Iraqi crude, shipped by pipeline to the Mediterranean, remained more expensive than comparable Gulf crudes.

Between 1958 and 1961, the newly established, radical Kassem regime, viewing the continuing marketing difficulties of Iraqi crude, became convinced that the IPC's continued restriction of Iraqi production was part of an American-British plot to bring down the government. Kassem believed that the American intervention in Lebanon in 1958, the continued British commercial presence in Iraq, and the continued unwillingness of the IPC to increase production (thus adding to Iraqi revenue) were part of a concerted plot to wreck the force of radical Arab nationalism. As a result, in April 1961 Kassem ordered a halt to all further exploration by the IPC.

In December of that same year, Iraq promulgated Public Law 89 which withdrew IPC's concession rights to all the areas in which it was not producing (99 percent of the concession area), including those areas such as the North Rumalia field where reserves were known to exist. In retaliation, IPC not only refused to recognize the validity of the government's actions but also threatened legal action against any company or country that obtained oil produced in any of the former IPC concession areas.

It is at this juncture that the actions of the United States Department of State severely escalated the crisis. Because it appeared that claims for compensation for expropriation were not justified, since no producing asset of the IPC had been involved, the Department of State's legal office indicated: "We have no firm legal basis for telling independent American companies—let alone foreign companies—to stay out of Iraq."[10] Nevertheless, despite this legal position, high-level officials of State, concerned that their silence on the issue might be viewed as tacit approval of the Iraqi unilateral act, intervened and formally advised smaller independent oil companies to stay out of Iraq and not attempt to gain access to the former IPC concession areas.[11]

As the crisis deepened, the Iraqi government, angered by the IPC's reactions (and those of the State Department), moved in 1964 to establish the Iraq National Oil Company (INOC) for the purpose of developing either on its own or in concert with other foreign companies those concession areas removed from the original IPC concession. In August 1967 the Iraqi government established Law 97, which created even more problems for the IPC. This law specifically barred the return of any area with known oil reserves to a private company. It also reaffirmed the right of the INOC to exploit and develop hydrocarbon resources throughout the country except for the small area reserved for the IPC under Law 80.

With the passage of Law 97, European companies (particularly

French) moved to gain access to the onetime IPC concession areas. The State Department protested vigorously to the French government, arguing that the acquisition by French interests of concession areas claimed by the IPC would create precedents elsewhere "which can weaken the security of Western oil rights and thereby adversely affect the national interests of France as well as of the United States."[12] The French government was deeply concerned over the intervention by the United States government on behalf of its companies. In the French view, the United States was far more concerned about American commercial interests than it was in maintaining the viability of western petroleum supplies.

From 1967 to 1972, amid growing domestic political turmoil in Iraq over the issue, the crisis continued unabated. Then on June 1, 1972, Iraq nationalized all IPC's properties except for the southern production of the Basrah Petroleum Company.

Whatever the historic accuracy of the Iraqi belief that the companies and the British and American governments acted in concert to retard the development of the Iraqi oil reserves, the fact remains that this perception became a deeply ingrained element in Iraqi politics (it was a factor in increasing Iraq's interest in a special relationship with the USSR, the first that country had obtained with an oil exporter in the Middle East) and has complicated relationships ever since, despite the intense Iraqi need for ever larger oil revenues.

I have only traced the outline of the oil disputes in these examples of Mexico, Venezuela, Iran, and Iraq which ended, invariably, in still further changes in the terms on which companies sought access to foreign supply. I have deliberately stressed the perspective of producer governments in these complicated and contentious matters, for it was the producer governments which took control. The erosion of the system whereby companies (backed by their own governments) had near total control also set the stage for the politicization of oil as producer governments moved in to exercise their newfound power. Nowhere was this more clearly demonstrated than in the events of 1973-1974.

3

The Challenge of Change
The Oil Importer & Oil Exporter Response

THE ECONOMIC dislocations caused by the unprecedented OPEC oil-price increases of 1973-1974 initiated a new era in international economic affairs. With OPEC's success in the "opening stage of a struggle for a new world order, a search for positions of strength in a global realignment, in which the weapons (backed, naturally, by the ultimate sanction of force) were food and fuel,"[1] the "crisis of interdependence"[2] had begun. Henceforth, intergovernmental political objectives would be inextricably mixed with the economic. In order to appreciate how complex these aspects have become, we need to remind ourselves of the various energy-related initiatives taken in the aftermath of that winter as the consequences of the use of oil as a political weapon and the ending of the concession system threw the international system into disarray.

THE OIL IMPORTERS

The economic crisis brought about by the embargo and the price increases created distortions in the economies of all oil-importing nations. In the case of the OECD nations, the crisis created tensions in the alliance system as the oil crisis presented widely varying challenges to its members.

The United States, Canada, Norway, and the United Kingdom possessed domestic energy resources (coal, North Sea oil, Alaska and Canadian oil and gas, tar sands) which could, in time, reduce these nations' dependence on foreign sources of oil. Other members, however, such as the Federal Republic of Germany, Japan, France, and Italy, saw little alternative but to continue to rely heavily on oil imports from the politically volatile Arab world, while at the same time pursuing a vigorous program to develop nuclear energy as an alternative in the generation of electric power. Because of their differences in oil-import vulnerability, OECD nations were unable, in the immediate aftermath of the embargo, to come to an agreement on the need for joint action. As a result, some countries (France, Japan) attempted to negotiate bilateral arrangements with key producers to ensure their continued access to vital energy supplies. For all its protestations against such moves, the United States succeeded in reaching agreements with Iran and Saudi Arabia which were widely believed to involve oil.

Those countries initially opposing joint consumer action (most notably France) did so out of fear that such joint action might be viewed by the producers as a hostile act, thus further jeopardizing the flow of oil. Belief that it was continued United States support for Israel in the 1973 Arab-Israeli war which had precipitated the oil crisis in the first place also contributed to the opposition.

THE SCRAMBLE FOR BILATERAL RELATIONSHIPS

While supply arrangements as a means of ensuring access to oil had existed before the embargo, OPEC's demonstration that oil prices and supply could be manipulated for overtly political reasons gave such arrangements a new sense of urgency. The motivation behind this effort for new bilateral arrangements was based on two untested assumptions: 1) that by creating special relationships with particular producers at all levels (economic, political, defense), oil supply security would become enhanced; and 2) that the creation of such relationships would help to restrain further price increases, or encourage petrodollar recycling, and thus help ease the balance-of-payments problems of the particular consumer nation enjoying the relationship.

In the wake of the embargo, attempts at government-to-government petroleum arrangements proliferated as governments came to realize that with the growing politicization of international petroleum trade, bargaining power had shifted away from the consumers (and their companies) to the producer governments.[3] Some major energy-consuming states feared that in the scramble for bilateral deals, failure to act would leave one's country on the sidelines in the evolving producer-consumer relationships governing future access to energy.

Once security of supply at a reasonable price became a prime component of government concern, issues of access to energy entered a new phase where more traditional commercial considerations were no longer the vital ingredient in oil negotiations. Nevertheless, it is not yet the case that the few arrangements concluded resulted in any particular advantage either as to volume, price, or some greater assurance of supply.

THE AMERICAN RESPONSE

While some governments of the industrialized world attempted to guarantee their access to vital oil reserves by negotiating even more complex supply arrangements (arms for oil, industrial goods for oil), the United States embarked on a policy designed ostensibly, and so described publicly, to lessen the chances for divisive competition. The objective of such a policy was to create a multilateral organization within which the national energy policies of the major industrial oil-importing countries could be coordinated. The United States hoped that by working to establish such a negotiating forum, it would assert its leadership of the industrialized world's response to the energy crisis, establish a unified front with which to negotiate with OPEC, control the character of the competition among the major consumer nations which only served to strengthen OPEC's bargaining position, and protect itself and others from a repetition of the 1973-1974 OAPEC embargo.

ALLIED COOPERATION

Although the institutional forum of the OECD had addressed energy problems for many years through its various energy-related committees—the Oil Committee, the Energy Committee, the Environment Committee, the Committee for Scientific and Technological Policy, and eventually the Nuclear Energy Committee—the United States was opposed to using the OECD organizational structure, since it believed that the OECD's past performance on energy, which reflected the lowest common denominator of interests, would make it impossible to achieve an effective consensus among the industrial nations. As a result, throughout 1974, the United States strove to establish a new and more effective institutional mechanism which could achieve such a consensus on unified consumer-country positions affecting supply. The United States believed that only when such agreement among the key OECD nations was achieved would it be possible to broaden oil discussions to include OPEC producers.

Despite Washington's intense diplomatic efforts, however, the Europeans, particularly the French, and the Japanese remained apprehensive of American motivations in establishing such a consumer-country forum and also remained concerned about producers' reactions to such a development. Given their much greater reliance on Arab oil, the rest of the OECD nations were reluctant to join an organization which the OPEC producers might view as confrontational. Nonetheless, despite the objections of French Foreign Minister Michel Jobert, the United States was able in February 1974 to convene the Washington Energy Conference which began the process of intraalliance cooperation on energy. As a result of the conference, all the participants except France agreed to form an Energy Coordinating Group to develop an "international action program to deal with the world energy situation on a cooperative basis."[4]

By June 1974 this preliminary consensus had evolved into an agreement in principle for an Integrated Emergency Program which combined provisions to share oil supplies and restrain demand during supply emergencies with longer-range efforts to conserve energy and to develop alternative energy sources. In August the mandate of the program was broadened to include provisions for obtaining detailed petroleum data from the international oil companies as well as a mechanism to improve cooperative efforts between oil-consuming and oil-producing countries. By November

19, 1974, consensus on these provisions had become sufficiently broadly based for sixteen nations to meet in Paris and to sign the agreement on an International Energy Agency (IEA) as the institutional mechanism for implementing the provisions of the program.[5]

Although a broadened consensus for a strong IEA evolved slowly during 1974-1975, the Europeans and Japanese remained skeptical of American intentions. Perhaps no event created such uneasiness as did the United States' move to create a special relationship with Saudi Arabia in the very midst of the negotiations for the creation of an IEA. The timing of the consummation of this special relationship during the delicate negotiations to set up the IEA sent Europe and Japan a warning signal: in the absence of an intraallied agreement on oil questions—or even with one concluded—the United States would act to protect its own vital interests in a relationship with the world's largest oil exporter. Despite suspicions of American motives on the part of the Europeans and the Japanese, however, the IEA emerged because the member nations understood the need to take cooperative measures to reduce their oil supply vulnerability.

While the United States worked to develop a strong and unified IEA, it remained a tenet of Washington's policy that only when the IEA established emergency oil-sharing and related mechanisms designed to protect the developed countries against another embargo and had embarked on plans to "transform the market conditions for OPEC oil" by reducing consumption and developing alternative energy sources, could the OECD nations enter into discussions with the OPEC oil producers. While the United States continued to hint at the establishment of some future producer-consumer forum, however, American policy throughout the winter of 1974-1975 remained more confrontational than otherwise, partly because an inflexible position on the part of producer governments did not seem to justify another tack. Throughout the winter of 1974-1975, and even into 1978, highest-level American officials openly discussed scenarios that could lead to United States military intervention in the Middle East to preserve access to oil while making other pronouncements which seemed to confirm that the main goal of American foreign economic and political policy was either to break OPEC or to penetrate its front through the exercise of United States influence over Saudi Arabia, the key OPEC member.[6]

THE NON-OIL LDC RESPONSE

While the economies of the oil-importing less-developed countries reeled under the impact of the fourfold rise in the price of oil during 1973-1974, OPEC's demonstrated success over the major industrialized countries showed the less-developed countries that the preponderance of world economic bargaining power had inexorably shifted from the developed world. This altered LDC perception was first revealed in April-May 1974 when at the Sixth Special Session of the United Nations General Assembly, the LDCs, led by Algeria, pushed through the assembly the Declaration of the Establishment of a New International Economic Order (NIEO).[7] Although the declaration primarily contained demands for international economic reforms (aid, trade, debt, technology transfer) which had been promulgated ever since the United Nations Conference on Trade and Development (UNCTAD) I in 1964, the fact that Algeria had taken the lead role in promoting it was important because it seemed to signify a willingness on the part of at least some members of OPEC to use the power of OPEC oil in advancing the interests of the less-developed world.

The unwillingness of the developed world to respond to the demands of the New International Economic Order (NIEO) led to a growing series of confrontations between the LDCs and the industrialized world. In December 1974 the United Nations General Assembly overwhelmingly endorsed the Charter of Economic Rights and Duties of States, encompassing both the demands of the NIEO and the Program of Action which had been under consideration for two years. Only six countries (all OECD members) led by the United States voted against adoption of the charter.

LDC manifestos calling for a fundamental restructuring of the world economic order were enunciated in Dakar in February 1975; at the Second General Conference of the United Nations Industrial Development Organization in March 1975; at the July 1975 meeting of the Organization of African Unity; and at the August 1975 meeting of the nonaligned countries in Lima. Although the "radicalism" of the NIEO's demands was much commented upon at the time, many of the demands, articulated by the Algerians, were designed to exploit the greater economic dependence of consumers on producers. The have-not nations of the world were demanding a change in the postwar economic status quo.

THE ALGIERS DECLARATION

Despite the opposition of the industrialized countries to third-world demands for a fundamental restructuring of the international economic order, in late 1974 there was the beginning, albeit embryonic, of a movement toward establishing a producer-consumer dialogue on energy. As a first step in paving the way for the commencement of producer-consumer energy negotiations, the United States and France, using an initiative originally proposed by Saudi Arabia, agreed in December 1974 in the Martinique Communique to promote a meeting between consumers (including representatives of the LDCs) and oil producers to attempt to work out joint positions among the major consumer countries in advance of the producer-consumer meeting. Simultaneously, the IEA agreed to coordinate its work in the spirit of the United States-French accord.

Following up on these initiatives, the oil, finance, and foreign ministers of OPEC, meeting in Algiers in January 1975, called on the OPEC countries to guarantee supplies to the oil-consuming nations until the oil consumers could introduce energy-saving programs and develop alternative energy sources. In addition, the Algerians called for a six-year transitional period during which OPEC would agree not to raise the real price of crude. In return for the implementation of these mechanisms, the industrial nations were to undertake commitments to assist LDC economies by maintaining high raw-material prices, by increasing their aid flow to the LDCs, and by giving OPEC greater participation in the institutions of the international monetary system (International Monetary Fund, the World Bank, General Agreement on Tariffs and Trade).[8] Although it was little noted at the time, the Algerian proposals represented a fundamental development in discussions revolving around energy considerations in that at least some OPEC members, notably Algeria and Venezuela, linked the fortunes of the developing nations with those of the western industrialized countries.

Ironically, given the lack of militancy of these proposals when viewed in conjunction with later OPEC demands, the United States failed to respond in part because the demands of the Algerian proposals posed a fundamental challenge to the American position that any discussion between consumers and producers should be confined to oil; that before any discussions could take place, the IEA member nations had to establish an emergency oil-sharing mechanism to protect against an OPEC embargo and move to reduce their consumption and to develop their alternative energy sources; that a dialogue with OPEC should be confined to discussing "an equitable price, market structure, and long-term economic relationship"; and that OPEC and the West should discuss the prospects of recycling OPEC petrodollars.[9]

A further reason for the United States government's reluctance to respond otherwise to the Algerian initiatives resulted from its concern that unless the major oil-consuming countries could agree not to allow imported OPEC oil to be sold in their domestic markets at prices which would make new energy sources noncompetitive (a minimum safeguard price), it could be risky or at least premature for oil consumers to enter into negotiations with the oil producers.[10] The attempts by the government to decouple the OPEC-LDC link by focusing its counterproposals on the need for OPEC-OECD cooperation in helping the OPEC nations recycle their petrodollar wealth seemed to the LDCs to be a further indication of the lack of American interest in their problems. This view was intensified in the aftermath of the Algiers Declaration. The United States, while proposing that OPEC should guarantee the West long-term crude supplies at fixed prices, gave only marginal attention to the interests of the LDCs, stating that both the developed countries and OPEC should join together to supply the LDCs with the appropriate technology and capital to foster their development.[11] The American failure to respond to these initiatives led the OPEC summit meeting in Algiers in March 1975 to issue the OPEC Solemn Declaration, which, with the demands made in Algiers in January and in the

Sixth Special Session, were designed to spell out clearly to the industrial world the integral link between OPEC and LDC interests.

THE APRIL 1975 PREPARATORY CONFERENCE

Although none of the consumer prerequisites (oil-sharing mechanism, minimum safeguard price) called for by Secretary of State Henry Kissinger had been met by April 1975, a Preparatory Meeting was held in Paris to set an agenda for an international conference between oil exporters and oil consumers to occur that summer. At the session, the embryonic alliance that had existed between OPEC and the less-developed countries emerged in broad daylight when OPEC, wary of the continued lack of serious discussion of their demands by the developed countries from April 1974, demanded that the talks on energy be extended to include other major agenda items of concern to the developing nations.

Because of the reluctance of the developed countries to place specific references to the nonenergy LDC and OPEC demands on the agenda, the Preparatory Meeting ended inconclusively. The lack of a response by the developed countries served in the LDC-OPEC view to confirm its perception that the developed countries had no intention of entering into serious negotiations with them on any changes in the international economic system.

The failure of the April preparatory session marked a key phase in North-South negotiations. Prior to the meeting, LDC demands might have been debated in the context of negotiations about such matters as oil payments and debt transfers, but afterwards, the North-South conflict turned into questions of principle and ideology. It was no longer a conflict about oil and oil prices but a conflict between two seemingly irreconcilable conceptions of what constitutes a just economic world order.[12] It would surely have to be based on something more than a share-the-wealth scheme or a simple reaffirmation of the existing system.

THE UNITED NATIONS GENERAL ASSEMBLY

Although the United States government began to modify its position on North-South issues in the late spring of 1975, it continued to refuse throughout the summer to reveal what issues it would raise at the September meeting of the Seventh Special Session of the General Assembly, a forum explicitly designed to discuss development and international economic cooperation. This stance was taken because the administration was enmeshed at that time in a major bureaucratic debate over how far the United States should go in meeting LDC demands. There were several signals, however, that the United States might actually be modifying its position. Secretary Kissinger announced the government's willingness to reconsider on a case-by-case basis its traditional opposition to commodity agreements; furthermore, the United States in May 1975 announced its support for the International Fund for Agricultural Development (IFAD). Although these two initiatives were heralded by some people as indicative of a more forthcoming policy by the American government, many LDCs remained skeptical because for twenty-five years the United States government, despite its ideological objections, had entered into commodity agreements on a case-by-case basis and because three months after the United States-IFAD commitment, no funds had been advanced.[13] The continued opposition of the United States government against giving raw materials and development problems equal standing with energy in negotiations plus continued opposition to oil-price indexation provided additional signals as to how far the United States was actually prepared to modify its position, or reargue the merits of its earlier stands.

Throughout the summer of 1975, the United States was under intense pressure from some industrial countries to enact policies more favorable to the LDCs. In this regard, it should be noted that one of the reasons for the European Community negotiations of the Lomé Convention in February 1975 with forty-six African, Caribbean, and Pacific countries was a desire to experiment with some of the LDC arguments, in order to help preserve an effective and stable relationship between those nations and Europe. Given their much greater vulnerability to a disruption of strategic commodity supplies, the European Community could be viewed as having gone beyond the American position. Moreover, there was some fear that if accommodations were not made, OPEC might use its oil to exact such concessions.

By September, the United States government had changed its position sufficiently to convince OPEC and the LDCs that their demand that energy problems be linked to the problems of the LDCs had been accepted. The change in the American position was launched by Kissinger's speeches before the United Nations Seventh Special Session in September 1975 and later in his December sixteenth speech at the opening of the Conference on International Economic Cooperation in Paris. In these speeches, he proposed a series of initiatives designed to alleviate the problems of the less-developed countries. Of these initiatives, the two most directly linked to energy were the proposals to create an International Industrialization Institute and an International Energy Institute. Whereas the International Industrialization Institute was designed to provide investment guarantees through some form of umbrella organization for private firms interested in LDC industrialization (including energy projects), the International Energy Institute as originally proposed was designed to develop energy assessments and strategies for particular LDCs, train energy officials and experts, provide technical assistance in energy resources exploration and development planning, and transfer and adapt existing energy technology to meet specific LDC energy requirements.

The importance of the Seventh Special Session is considerable. The most vital and interesting aspect of the session was the reemergence of the moderates within the Group of 77 (LDCs). During the meeting, this group of nations, tired of empty political rhetoric but anxious to achieve a resolution on concrete international economic reforms acceptable to the industrialized countries, moved to ensure that they would dominate the southern bloc of nations. The significant actions taken by these moderates included the nomination of an OECD delegate to serve as chairman of the ad hoc committee given the task of drawing up a draft resolution acceptable to both the northern and southern camps; a willingness to accept Kissinger's proposals at face value and to suspend sharp criticism of them despite the fact that the speech contained few concrete concessions to their demands; and agreement by the Group of 77 to a consensus resolution which was a watered-down statement of their views as expressed since the Sixth Special Session.[14] From the American side, the government acknowledged in effect that the southern bloc's demand for negotiations was in itself a legitimate request and called for the beginning of negotiations with the LDCs in a series of international forums.

The interesting part of the Seventh Special Session's deliberations was the degree to which the small shift in the American position, amounting to insignificant financial concessions, was seized upon by the moderate members of the Group of 77 to reduce the influence of the radicals. This shift seems to indicate that as long as the United States systematically opposed all the demands of the Group of 77 from April 1974 to September 1975, the moderates within the group were eclipsed by the radicals who exercised a degree of influence far disproportionate to their numbers. Once the moderates were given an opening by the United States, however, they could move to assert their influence.[15]

The final resolution adopted by the Seventh Special Session called for a number of steps, every one of which contained implications for energy:

(1) expanding and diversifying trade, improving productivity and increasing export earnings of the developing countries; (2) securing stable, remunerative and equitable prices for exported raw materials, and protecting their purchasing power; (3) reducing or removing tariff and non-tariff barriers affecting the less-developed countries' exports; (4) increasing the volume and improving the terms of development assistance; (5) achieving the official development assistance target of 0.7 percent of donor countries' GNP by the end of this decade; (6) increasing the LDCs access on favorable terms to world capital markets; (7) relieving debt burdens of the most seriously affected countries; (8) giving the Third World a greater voice in the management of international financial institutions and larger access to their resources; (9) increasing international control and surveillance over the creation and equitable distribution of world liquidity; and (10) facilitating the process of industrialization in the developing world.[16]

THE JAMAICA ACCORD

The spirit generated by the Seventh Special Session led in January 1976 to the Jamaica Accord which was designed to reform the international monetary system. Under the terms of the reform package which the developing nations received were an increase in their International Monetary Fund quotas from about 7.5 billion units of Special Drawing Rights to 12 billion, an immediate 45 percent enlargement of their use of fund resources; a trust fund through which they could receive the profits from the fund's sale of gold; and a reduction in the time for further review of the fund's quotas from five to three years.[17] Although these reforms offered the LDCs some positive benefits, the unwillingness of the International Monetary Fund to establish a link between access to the Special Drawing Rights and development finance problems created acute problems for the LDCs.

With the escalation of the total LDC external public debt from $108.7 billion in 1974 to $132 billion in 1975 and $157.1 billion in 1976, the LDCs' development prospects were severely diminished. The fact that much of this rise in the level of debt was due to the rise in the price of imported oil posed severe problems for the LDCs (and for OPEC members as well). The failure of countries' acting through international institutions to enact measures that would have effectively helped to alleviate the problem forced many LDCs to cut their oil and fertilizer imports which led to a decline in both food and industrial production. While the volume of oil involved was not significant, OPEC nations could only acknowledge their own great responsibility in bringing such a situation to pass, while, at the same time, attempting to foster the LDC cause.

CONFERENCE ON INTERNATIONAL ECONOMIC COOPERATION

With the launching of the Conference on International Economic Cooperation (CIEC) in Paris in December 1975, the North-South dialogue, as the various OECD-OPEC-LDC issues are now described (and simplifying what is actually a complex set of interests more than blocs), was divided into four commissions: Energy, Raw Materials, Development, and Finance. Although it was initially hoped that the various commissions would achieve tangible results, and despite subsequent alterations in views in other forums, the motivations of the participants ensured this would not occur. Because the United States believed that a discussion of non-oil-related issues in the CIEC forum would divert the conference from meaningful progress in energy, the government, supported by other industrial members, consistently used the Financial, Raw Materials, and Development commissions as debating forums to gain time in the Energy Commission. As a result of these delaying actions, the United States government was not, in the LDC view, an honest participant in CIEC. Thus many LDCs never took seriously the various energy-related initiatives, proposed by Secretary Kissinger in CIEC and elsewhere, to deal with the energy problems of the less-developed countries.

AMERICAN INITIATIVES IN THE CIEC

A principal difficulty with the proposals of the International Energy Institute, the International Industrialization Institute, and the International Resources Bank (IRB)[18] was that they were launched with little or no concrete thought given as to what these institutions were designed to do and how they might accomplish their objectives. There was never a systematic attempt to coordinate either these various American initiatives among themselves or to coordinate them with similar activities being conducted by international institutions such as the United Nations Industrial Development Organization, the United Nations Secretariat (Energy Branch), the International Atomic Energy Agency, the World Bank, and the International Monetary Fund (Energy Facility). The government often did not discuss various initiatives with the Europeans or Japanese before they were floated in CIEC (or elsewhere). As a result, other consumer countries were ill-prepared to support United States initiatives.

The chief problems hindering the achievement of concrete results in CIEC, in addition to the industrial states' opposition to discussing non-energy-related issues, were the unwillingness of the developed countries to abandon other international forums discussing North-South issues where their influence was greater, the economic recession in the industrialized nations which made them suspicious of enacting reforms that might threaten economic recovery or rekindle inflationary pressures, opposition to LDC-proposed commodity agreements seeming to do little to ensure supply while raising prices, and unwillingness to appear to give into the collective pressure of OPEC.

A major American initiative—the International Energy Institute (IEI)—deserves fuller discussion. When this initiative was launched, a principal objective of United States policy was to keep the IEI from having any but minimal ties to the United Nations. This strategy was based on the belief that the militant LDCs controlled the United Nations and hence would control the IEI if strong links were established between it and the United Nations. The weakness of this proposition, since the United States government had first proposed the IEI before the General Assembly, should have been apparent, particularly since the inadequate funding of the IEI ($6 million) had spurred the LDCs to establish a much better-funded organization within the United Nations Secretariat. Clearly, the United States wanted to keep the IEI in a forum where its influence would be strong.

The IEI concept was a good one; it could have been a forceful initiative and deserves a more effective reintroduction. It failed, however, because those who proposed it never addressed these important questions:

• What did the United States want the IEI to be?

• What was the United States willing to give to make the IEI an effective initiative?

• Would the limited budget of the IEI allow the serious consideration of the energy problems of over one hundred oil-poor LDCs?

• What role was envisaged for the IEI in the process of technology transfer to the LDCs?

• Was the IEI designed to transfer technology to the LDCs, or was it designed to provide technical assistance?

• Would the United States make its technology available to LDCs?

• What were the potential legal problems with regard to such questions as patent applicability?

• Did the United States envisage the IEI as a technology-transfer vehicle or as a mechanism for financing international resource development?

• Since the scope of IEI funding excluded resource development, what help could the IEI provide to an LDC that could not afford to develop its own resources?

• What could the IEI do for the LDCs that multinational corporations could not do if terms of access and repatriation of profits were reasonable?

• Would the IEI initiative have been more acceptable to the LDCs if it were tied to a respected nonpolitical institution such as the World Bank?

If the IEI proposal had been offered as part of a broader economic package, it might have generated a positive LDC-OPEC response. Standing alone, however, it could only fuel the belief of the more radical LDCs and OPEC states (Algeria, for example) that the IEI was a ploy, designed either to prevent CIEC from discussing the more important LDC issues in the trade, aid, debt, and technology-transfer areas or, alternatively, as a means for the developed countries to perpetuate economic neocolonialism.

THE LEGACY OF CIEC

With the collapse of the CIEC negotiations in June 1977, following a year and one-half of inconclusive negotiations, it is apparent that there remains a wide gap in the perceptions of the LDCs, OPEC, and the industrial states on what reforms (if any) need to be made in the international economic system. If the continuing "crisis of interdependence" is to be depoliticized, then some formal agreement securing access to energy and other raw materials must be achieved, and this can probably be accomplished more surely as other issues affecting commodity trade become resolved. Oil is different from other raw materials largely because it has no ready substitute and it is difficult and expensive to stock. But it is not unique.

While some industrial states have signaled a willingness to accommodate some of the LDC and OPEC demands through the enactment of specific measures, such as the Lomé Convention, the LDCs are increasingly concerned about what they have really gained from supporting OPEC vis-à-vis the developed world, or in accepting OPEC sponsorship. There is a growing perception among the LDCs that OPEC members are much more concerned about their own immediate problems (security of financial assets from the effects of inflation and/or currency devaluation, investment guarantees against political risk, a greater voice in international monetary affairs, and more substantial western and Japanese assistance in their industrialization and economic diversification programs) than they are in easing pressures on the economies of the less-developed countries. Indeed, by mid-1978, the failure of OPEC to use oil to wrest vital concessions from the industrialized world which would directly benefit the LDCs has led some LDCs to the conclusion that in the absence of a two-level oil-pricing system favorable to them, higher OPEC oil prices are at least equally harmful, if not more destructive, to their economies than they are to the economies of the industrialized countries. Increasingly, these countries are coming to the realization that they must seek out new arrangements which will enhance their economic and political security.

It is ironic that as particular LDCs have begun to look for new relationships to protect their vital interests, the industrialized world and the conservative members of OPEC have begun to realize that many of the world's economic issues are indeed inexorably linked (oil, debt servicing) and that their continued failure to address these issues may only be at their own expense or even peril.

Although the LDCs have ample justification to believe that OPEC has been less than forthright in such matters as providing adequate increased aid flows and debt moratoria, there can be little doubt that the move by Algeria to couple the oil issue with LDC-related issues was of some benefit to the LDCs. Until the establishment of the OPEC-LDC link, the developed nations had neglected the more serious LDC-related issues—debt, trade liberalization for LDC processed and semiprocessed goods. Only when the West perceived that its reluctance to consider these issues could provoke OPEC (especially Saudi Arabia) did the West begin discussions on commodity price stabilization schemes and related matters.

While the "oil link" appeared to be a relatively inexpensive way for OPEC to demonstrate its solidarity with the LDCs—born out of a common perception of historical grievance and providing a means of securing LDC diplomatic support on the Arab-Israeli issue, a means of increasing the costs to the developed countries of any precipitous military action under consideration, and/or a way to increase the role and leadership of individual OPEC members in a united LDC bloc and in specific international forums—the failure of OPEC and/or the industrialized world to deliver a meaningful reform package to the LDCs has disturbing implications. More specifically, failure of the CIEC in June 1977 to establish a new international body where oil exporters and oil importers could continue to consult on questions relating to energy access is nearly inexplicable, particularly in light of the observation that both the world's largest consumer—the United States—and the world's largest exporter—Saudi Arabia—apparently favored the creation of such a mechanism. Although the developed nations made some last-minute concessions to the LDCs—$1 billion extra aid and a plan to begin negotiations on stabilizing raw material prices later in the year—these proposals were rejected by the LDCs as desperate measures to obtain LDC support for establishing a new energy consultative mechanism.

In the view of the LDCs, vigorously supported by Iraq, Venezuela, and Algeria, the failure of the industrialized countries to agree on a wholesale indexing of the prices of raw commodity exports of the LDCs to the prices of imported industrial goods from the West and the failure of developed nations to make any positive concessions in the area of debt, the transfer of technology, and reducing barriers to exports from poor countries demonstrated that the North had no serious interest in addressing problems outside of energy. Nevertheless, despite CIEC's failure, the LDCs realize that it is in their interest to keep an energy consultative mechanism alive, because without such an institutional device future oil negotiations between the major consuming and producing nations will most likely occur in an environment in which oil is no longer linked with larger LDC interests and, indeed, their own desperate concern over oil prices and supply would be discussed and perhaps agreed upon without their presence.

It is because of the LDCs' potential exclusion from the bargaining process that the failure of the CIEC and the growing divisions in all camps have disturbing implications for future oil negotiations. To the extent that a group consensus, albeit presently min-

imal, is replaced by states' seeking new oil supply and p
rangements, an intense competition and even conflict over a<
oil could be our lot.

WORLD OIL IN THE AFTERMATH

It has been less than a decade since the oil producing and exporting
nations took control over the disposition of their oil resources from
the international oil companies. The speed with which this has been
accomplished, by states acting unilaterally, individually, and col-
lectively, has been such as to make it difficult even to hazard a
guess as to how the present period, which is still one of transition,
will evolve. As is the case in the international oil industry, modifi-
cations of past roles are likely to be more the case, imposed by con-
suming-importing and producing-exporting governments, than
sudden, more revolutionary steps eradicating them in favor of national instruments. The international movement of oil remains a
complex process which requires international mechanisms.

In these recent years, however, as government has steadily diminished the decision-making power of the international oil company, the resultant politicization of oil, born of nationalism and the
achievement of political independence since 1945, has profoundly
altered the terms on which oil will be supplied. Yet it is far from being the case that producers control the flow. Not only are most of
them limited in their power to affect supply because of their ever-
increasing need for revenue and their absolute dependence upon the
huge volume markets of the industrial world, they are also aware
that irresponsible moves in price and levels of production can
create great difficulties for importing nations, difficulties that can
run the risk of a military reaction. There is fuller appreciation of
the point that producers and consumers are mutually dependent
and that vital interests on the part of each group must be taken into
account.

It has not yet been possible to work out a mechanism or create
the inducements that will cause a producer, especially Saudi Arabia
and the United Arab Emirates, to pump oil at a volume that generates revenue surplus to their needs; nor has Norway been per-

suaded to do so either. The same problem may arise with others, namely, how to obtain levels of production adequate to meet world needs in the period of transition from oil to other forms of energy.

One effort has ended in disappointment: the inconclusive ending of the Conference on International Economic Cooperation. The various United Nations forums continue to examine and debate the issues in a new economic order—issues of price, volumes, stable revenue, technology transfer, each of which bears upon access to energy as well as to food and minerals.

Mankind may have other opportunities to begin again to find a process by which governments can discuss their estimates of need against probable levels of available oil, where considerations of price can be examined in the context of general world economic conditions, and where the particular needs of non-oil LDCs can be discussed. Whether these lead to a kind of commodity agreement for oil or to a less formal set of understandings, whose essence is a deeper appreciation of the ramifications of changes in oil terms and volumes, the crucial point is that these considerations must be isolated to the fullest extent possible from international politics if there are to be durable and equitable undertakings regarding adequate supply at prices related to nations' capacities to pay. If this degree of understanding as to the ramifications of our mutual dependence is not achieved, then a scramble for oil is well-nigh certain on a scale unprecedented in petroleum history. With great interests at stake, governments will be expected to use every instrument available to them including force. Moreover, it is no exaggeration to state that such a competition for oil in world trade can rend the North Atlantic Treaty Organization and wreck the Common Market. Access to energy in the form of oil is that essential. Anticipating a period of probable shortage in the mid- to later 1980s, governments must consider steps soon to assure their needs are met.

There were periods of surplus in earlier years when the companies were the managers of a system that sought continually to balance demand and supply. As the system evolved out of its original concession mold, however, governments moved in, the world of politics intruded, and the question of access became infinitely more complicated.

There may be reason to hope that in time the United Nations Secretariat's efforts to assist LDCs in their negotiations over energy development and in energy technology itself will allow the LDCs to obtain their social objectives and to obtain increasing amounts of petroleum to fuel their economies. Another hopeful initiative is the World Bank's new energy program. It is the intention of the bank to make available money, technical advice, and negotiating counsel to LDCs to determine whether they may have within their borders at least enough oil to reduce or perhaps to eliminate their own oil imports. It has been argued that the large international oil companies cannot afford to develop resources whose magnitude indicates they could meet only the domestic requirements of the LDC. The larger companies must usually make their investment decisions on the prospect that a developed find would fit their worldwide supply network. Hence, the bank intends to see if there is a useful role in meeting the lower-level requirements of a particular LDC. This involvement of leading multilateral, official, international lending institutions in an area previously dominated by private or single-government initiatives will prove the value of its undertaking only with years of application; nevertheless, it is an exceedingly interesting initiative from the viewpoint of many LDCs for access to oil in world trade, where by and large they cannot hope to compete, least of all in a time of tighter supply.

4

The Future of the International Oil Industry

THE evolving relationships between the international oil industry and the governments of the oil-consuming and oil-producing states could be of fundamental consequence to those people who seek to assess the future of international energy supply. It was the integrated operations of the oil majors that created the international oil system which moves a billion barrels of oil every day from one location to another—an extraordinary accomplishment. The efficiency of the system is an essential ingredient in assured supply.

DEVELOPMENT OF THE INTERNATIONAL OIL INDUSTRY

During the first half of the twentieth century, the close coincidence of interests between the major international oil companies (IOC) and their parent countries led to situations in oil-producing countries where the activities of the companies were often regarded as commercial instruments of the overseas political and economic objectives of the major oil-consuming states and thus, more often than not, viewed as part of a colonial system. As a result, relationships between the IOCs and the oil-producing states were often influenced by considerations of nationalism, colonialism, and differing interpretations of commercial and treaty law, as well as by issues of national sovereignty arising out of the international direction and global operations of the companies.

The international oil companies applied their many assets to the exploitation of oil—generating capital, using unusual management abilities, applying technological knowledge in the production, transportation, refining, and marketing of crude, and developing the integrated worldwide logistics system. The financial benefits of the IOCs' operations were largely received by the companies (and consumer governments in the form of taxes on oil products).

Although the link between the interests of the major consuming states and the international oil companies was maintained for many years, the Mexican petroleum nationalization of 1938 demonstrated for the first time that the power of the IOCs, even when backed by home governments, could be challenged. It is also well worth emphasizing that in that prolonged confrontation, the use of armed force was not resorted to and has not been used in any subsequent confrontation between oil producers and companies although financial, economic, and political pressures to sustain a concession or to maintain a contract were continuously applied. The use of force in the traditional military sense has not, to date, been engaged.

In the wake of the Mexican actions, some of the IOCs began to change their relations with the governments of producing countries by instituting programs for the technical training of host-country nationals in aspects of the petroleum business, by building schools and housing, and by providing medical care. Although these programs were established not out of simple altruism but from a desire to protect the security of their investments, these initiatives did connote an understanding on the part of the companies that in a world of rising nationalism, some change in producer government/company relations was necessary.

During the early 1950s, however, the domination of international oil by the majors was challenged on several fronts. In the commercial arena, the majors found themselves in growing competition with established national oil companies such as Ente Nationale Idrocarburi and Compagnie Française des Petroles and a number of smaller independents which increasingly entered the international market and whose interests were often limited to one or several producing countries and usually were involved in less than the entire set of functions in the system characteristic of the giant companies. This distinction between the interests of the majors and the independents became a matter of great tactical significance as the oil industry and producer governments squared off in their confrontation. The independent could have most of its overseas interests tied up with one country and could offer more or would concede more quickly to demands for change. On the producer front, the companies found their position increasingly attacked by nationalistic parties which emerged in the postwar years. Because these nationalists regarded the companies as part of their earlier subordinate status, there were ever-burgeoning demands for greater control and regulation of IOC activities.

The Iranian oil crisis of 1951-1954 marked a watershed in the relations between oil-producing governments and the international oil companies in that it demonstrated to the producing countries that until they controlled pricing, production, and the logistics system, they would never achieve their full sovereignty; the foreign oil companies' control over the exploitation of a domestic resource became a major target for political activists.

TOWARD GREATER GOVERNMENT CONTROL

With the rising tide of nationalism in the third world during the 1950s and 1960s, producing governments began to enact measures to acquire a greater share of the revenue derived from oil. Although during much of the 1950s the expansion in world petroleum demand allowed the companies to market most of the oil produced, thus meeting the demands for ever-greater volumes of crude, the growing competitive challenge from both state-owned companies and the independents had by late 1959 contributed to a situation in which there was a substantial surplus of crude oil in the world market. Control of the international oil market was passing from the IOCs, beginning first with the new commercial entrants and later with the producer governments. To reduce the surplus, the international oil companies in 1959-1960 twice lowered the posted price of crude, actions that precipitated the formation of OPEC.

Although the oil-producing governments were still slow to realize the extent of their bargaining power over the companies, the 1960s did witness a number of changes in company-government relations. Special allowances for the sale of crude oil were phased out; the practice of treating royalty payments as a deduction from the oil company's local tax obligation was terminated; oil-company assistance to the local economies was expanded; and company and government representatives met increasingly to negotiate mutually agreed-upon improvements in oil terms.

As world demand for oil throughout the 1960s mushroomed, governments in both oil-producing and oil-consuming states moved to enact national oil policies which varied greatly in scope from country to country. The motivations behind increased governmental intervention arose from an amalgam of concerns including the critical role that petroleum played in national security and economic growth, balance-of-payment problems arising from escalating levels of oil imports, a mistrust of foreign corporations, and exploitations by foreign interests.[1] According to Neil Jacoby, the postwar oil-policy goals of foreign countries were determined by the interaction of three main variables: 1) the relation between a nation's indigenous production and its consumption of oil; 2) the extent of the ownership of foreign oil by its nationals; and 3) the role (declining) of domestic coal as a competitive energy source.[2] Thus, while the oil-surplus countries sought higher oil prices and expanding markets, oil-deficient countries with large coal deposits but without large interests in foreign oil sought low-priced oil while some countries extended their foreign petroleum investments. At the same time, they placed excise taxes on imported petroleum products for general revenue and also to lessen the economic impact of the transition to oil on their domestic coal industries.[3]

Oil-deficient countries that possessed neither sizable domestic coal reserves nor foreign oil holdings (Italy, Japan) pursued policies of seeking low-priced oil. Others, such as the United States, Great Britain, and France, which possessed significant foreign oil interests but also had large domestic coal resources, enacted measures to accommodate conflicting energy interests. But generally, everywhere, the objective of using the cheaper energy source represented then by oil overcame prudence and led rapidly to significant dependence upon imported oil.

Governments pursued a variety of policies aimed either at reducing the power of the IOCs or in obtaining a piece of the action long withheld by the IOCs: they expropriated and nationalized private oil properties, moved to regulate private oil development and exploration activities, enacted measures governing the refining and marketing of oil products, enforced antimonopoly laws, raised or reduced levels of oil production, and increased government regulation of petroleum pricing and product decisions.[4] The involvement by governments in oil was increasing and nearly universal. By the early 1960s, as it became apparent that spot shortages might occur if the demand for oil were to continue to rise, oil-producing states realized that oil could have been underpriced both in regard to the producers' share and in regard to alternative energy fuels. This belief was confirmed for them by the fact that throughout the 1960s and on through 1970 crude oil in world markets was priced lower than in the United States quota-protected market.

THE LIBYAN CONFRONTATION OF 1970

The importance of Libyan oil production in world trade during the 1960s was unusual in two respects: the rapidity of its growth and the role that independent companies played in its development. While not to be considered among the largest of the oil exporters, Libya became the tactical leader in initiating the greatest changes in the role and power of the international oil companies. From a production of only 182,000 B/D in 1962, Libyan production skyrocketed to 3,318,000 B/D in 1970, making it the fourth-largest OPEC producer, behind Saudi Arabia, Iran, and Venezuela.[5]

The main factors favoring the rapid development of Libyan crude were its low sulfur content, which made it an increasingly desirable fuel in the pollution-conscious nations of the industrial world, and its close proximity to European markets, which gave it an important transportation differential over long-haul crude from the Persian Gulf. The Libyan oil industry was also unique in that from 1955 to 1969, the government had welcomed independent companies, which, lacking concessions elsewhere in the world, would succeed or fail on the basis of their interest in Libya. The government believed that by favoring the independents it would avoid the problem of the majors' possibly reducing their output in Libya in order to avoid a world surplus, thus protecting their larger interests in the Gulf.[6]

The rapid rise in Libyan production confronted the majors with a dilemma. While their own concession areas in Libya were highly profitable, the majors realized that if the Libyan output of the independents continued to rise it might begin to make inroads on crude sales out of the Gulf. However, realizing that the quality advantages of Libyan crude could lead to a reduction in Gulf output, thus reducing the revenue accruing to those governments, the majors feared also that the oil producers would blame them for any decline in their production or sales. Because their production interests in the Gulf were some twenty times greater than their production in Libya, the majors were justifiably anxious not to antagonize the Gulf producers. At the same time, the majors realized that if they abandoned Libya to protect their Gulf interests, their concessions would be turned over to the independents, thus expanding the amount of oil on the world market not controlled by the majors[7] and leading probably to accelerating demands for change in the IOC-producer country relationship.

With the overthrow of King Idris of Libya in 1969, the radical regime of Mu'ammar Qaddafi, realizing the predicament of the companies, cleverly isolated one major (Exxon) and one independent (Occidental) and demanded new price negotiations. Because Occidental was almost totally dependent on Libyan production to service its European markets, Libya was able to force it substantially to increase Libya's share of the petroleum-generated revenue. Although Occidental tried to enlist Exxon's support to oppose Libya's demand, Exxon's reluctance is thought to have helped ensure a Libyan victory.[8]

Libya's success in isolating the companies on the basis of their different interests began a process of leapfrogging, where no sooner was a new agreement concluded in one producing country than its stiffer terms were used against the companies by another producing country. This process, which was supposed to be ended by the Teheran Agreement but was not, led to a succession of price re-negotiations and other adjustments, including a greater share of government participation. From 1971 to 1974 the participation interests of the producing governments rose from 25 percent to 60 percent and were still climbing.

The Libyan crisis of 1970 and the subsequent Teheran Agreement of 1971 marked another watershed in relations between oil companies and producer governments. The terms of access were shifting rapidly, dramatically, and irrevocably.

THE KUWAIT DECLARATION OF OCTOBER 1973

Despite the significant changes brought about by the Teheran-Tripoli-Geneva Agreement of 1971-1972, there was perhaps no more critical breach in the relationships between the companies and the oil-producing governments than that implied by the Kuwait Declaration of October 1973. This declaration occurred against the backdrop of the Arab-Israeli war which commenced on October 6. It announced the doubling of the posted price of crude oil on a take-it-or-leave-it basis instituted unilaterally by the oil-producing countries. On the same day that the Kuwait Declaration was announced, the Arab oil producers (OAPEC), in response to the United States' announcement of its intention to resupply Israel with military equipment, issued a call for the imposition of an oil embargo. The OAPEC embargo placed all those companies receiving crude oil from the Arab oil producers in an extremely difficult position; the embargo directive placed the companies right in the middle of the political conflict between the oil-producing and oil-consuming nations.

Although the companies effectively and equitably used their control of the international logistics system to divert oil supplies to alleviate the full, adverse effects of the embargo, the result of their actions was sometimes to estrange them not only from the Arab oil producers but also from the public in some of the oil-consuming countries[9] that did not believe equitable sharing of the shortage was in fact occurring.

INDUSTRY-PRODUCER GOVERNMENT RELATIONS SINCE 1973

Little is left today of the agreements which once defined the international oil companies' relationships with producing countries. The preeminent interest of the company has been replaced by that of the country. Nevertheless, the power of producing governments is not absolute. Price and levels of production are now set by government, but general market conditions continue to be influential. Rivalries between producer governments and a host of other considerations set limits to their decisions. More readily apparent, however, is the continuing need for the technical knowledge and accumulated experience of the industry in exploration, field development, and management of the international oil logistics system. While the experience of producing governments is growing, and the number of companies technically competent in one or another aspect of oil operations increases, it is probable that for many years to come the international oil industry will be indispensable to the efficient supply of oil in world trade. A major threat to their ability to perform comes from the possibility that an explosion of an issue in the Middle East would lead to an intense political reaction against the companies. Even so, once time passed, the capabilities presently found largely only in the industry would be as essential as before.

Of greater consequence now may be the actions taken by oil-importing governments to press for greater participation in IOC decisions or to substitute national companies with a preferential position in imported supply for the international system of the IOCs. These governments would take such action in a period of short supply to obtain some special advantage over another importer. While the International Energy Agency has emergency plans to share equitably the oil that may still be in world trade, it was not designed to cope with chronic shortages.

The growth of national oil companies, many with overseas interests of their own, will have many ramifications, but none will be more consequential than if governments influence the decisions of these entities, thus greatly increasing the possibility of still further emphasis on the politics of oil.

The risk of divisive competition between governments demonstrates clearly the need for a process where discussions between the producers and consumers of oil as to available supply, measured against anticipated demand, may serve many purposes, including maintaining the efficiency of the world logistics system—a matter of great significance to the exporters and importers—which is an essential ingredient in any discussion of access to energy, especially throughout the balance of the petroleum era.

The transition from oil to alternative forms of energy will be accomplished neither easily nor soon. The issues involved in oil inform us not only that questions of exploitation which will be with us after 2000 are complex but that they are probably already to be found in the expansion of the use of nuclear power. It is to these nuclear concerns that energy-deficient nations now turn; their early resolution will help significantly to shape the world's longer-term energy future.

5

Nuclear Energy
To 2000 and Beyond

ONE of the major consequences of both the 1973-1974 OAPEC oil embargo and the escalation in world petroleum prices was the growing recognition by all oil-importing countries of the need to reduce the level of their dependence on high-priced supply-disruptable oil. The use of nuclear power appeared to be an attractive option both because the theoretical technology was well proved and because a great deal of scientific work and practical experience had been accumulated. Four years after the embargo, however, the enthusiasm for the timely development of nuclear power has waned. Instead, the debate over the future of nuclear power has, increasingly, polarized into an argument between those who advocate the rapid expansion of nuclear power programs as the only major short-term solution for alleviating the world's energy crisis and those who believe that the potential costs associated with rapid nuclear development far outweigh the benefits.

The controversy over the costs/benefits of nuclear power has been fueled by a combination of issues, among them: 1) the rapidly escalating costs of nuclear plant construction and uranium fuel supplies which may remove nuclear power from competition with other alternative fuels; 2) the extent and availability of nuclear fuel supplies; 3) the health and environmental aspects of accelerated nuclear-power generation, especially those associated with nuclear-waste storage and the advent of the plutonium economy; 4) the potential safety of nuclear power in comparison to other alternative fuels such as coal; 5) the relationship between the worldwide growth of nuclear power and the potential for the proliferation of nuclear weapons; and 6) the danger posed to the international community either by terrorist sabotage of nuclear installations or by terrorist acquisition of a nuclear explosive device.[1] While each of these issues is itself subject to debate, the controversy over the future of nuclear power rests on the conflict between those nations who view nuclear power as dangerous, because of its weapons potential, and those countries who, while accepting the possibility of weapons proliferation, see nuclear power as the only alternative to dependence on imported energy sources.

Unless one assumes oil will always be found, the need for an alternative to oil is unarguable. Some people think that with a highly disciplined energy program we might be able to move from the oil era into an energy age in which renewable energy sources would allow us to avoid dependence on nuclear power. Such convictions are usually based on the premise that the industrial world can commit itself to such an energy effort—drastic enough in its implications of low economic growth—and that energy research and development will produce the new energy sources on a massive scale, and in time. Failure in such an effort would result in an energy shortage of catastrophic proportions. Those who believe we must take advantage of nuclear energy do not discount the danger of weapons proliferation, nor minimize the risk of sabotage or terrorist acts, nor minimize the continuing safety and waste disposal issues. But these problems can be dealt with in a reasoned manner, if never wholly resolved to any thoughtful person's satisfaction.

NUCLEAR ENERGY SINCE THE EMBARGO

In the aftermath of the OAPEC embargo of 1973-1974, one view was that, within a relatively short span of time, the primacy of petroleum as a fuel would come to an end, to be replaced by uranium, coal, and other alternative energy sources. As research into alternative fuels commenced, however, it was soon apparent that constraints on the supply and availability of uranium reserves might soon be encountered, and that the costs of alternative fuels would continue to rise at a rate faster than that for imported petroleum, thus effectively retarding their development.

Because it would generate more energy than consumed, therefore reducing world concern over the availability of natural uranium reserves, the development of the fast-breeder reactor was heralded as a potential panacea.[2] The development of the breeder reactor, however, was tied to the "Faustian bargain"[3] of developing an economy which with all its evident energy advantages could also pose grave environmental, health, and national security dangers. Although many of these dangers exist even with the conventional generation of reactors now producing commercial electric power, some observers feel that the development of the plutonium economy increases these dangers considerably. Unlike energy sources, the danger of nuclear weapons proliferation, arising from a rapid expansion in the world's use of nuclear power, raised fundamental issues that cannot be ignored in any discussion of the pros and cons of nuclear-power development.

It is this qualitative difference more than any other that has led to a growing politicization of the nuclear issue in all developed societies. Indeed, the danger posed by the "Faustian bargain" was the fundamental catalyst in causing President Jimmy Carter to launch his program both to impose tighter restrictions on the international exchange of nuclear-reprocessing technology and to curtail substantially the development of the Clinch River commercial prototype of the United States fast-breeder reactor whose full contribution to nuclear technology could be as distant as the next generation. President Carter's domestic nuclear program, molded in the belief that the dangers of some aspects of nuclear power far outweigh its advantages, raises fundamental questions when viewed in the context of the international arena. This policy will affect the ability of certain states to ensure access to energy supplies, which could in turn lead to the possibility of a small group of nuclear-technology-supplier nations banding together and making decisions that would affect the entire international community. It is of paramount importance to note that the resolution of these "nuclear" issues during the remainder of the decade will vitally affect the geopolitics of energy well into the next century.[4]

In order to evaluate the international nuclear debate, it is first necessary to understand the underlying technology which is involved in the production of nuclear power for both commercial and military purposes.[5] There is a complex set of factors and choices which must be resolved if the promise—or the specter—of nuclear power is to be the end result.

NUCLEAR ENERGY AND NUCLEAR WEAPONS

Because fissionable material is the key ingredient in the development of nuclear weapons as well as in the generation of commercial nuclear power, it is impossible to discuss the question of future access to commercial nuclear power without simultaneously addressing the major military and environmental implications of such developments.[6]

When a commercial nuclear reactor produces nuclear energy, plutonium, from which nuclear weapons can be manufactured, is a by-product of the reaction. Irradiated plutonium (that produced in a nuclear reaction), however, cannot be readily converted into a bomb until it is separated and recovered in a nuclear-reprocessing plant. It is because of this reprocessing stage in the nuclear fuel cycle of commercial power reactors where nuclear material directly usable for weapons is produced during normal operations that the current debate arises on curtailing the export of such technology (German-Brazilian and French-Pakistani deals).

During most of the last thirty years concern about nuclear proliferation has centered on the dangers arising from the production of plutonium as a by-product of a nuclear-fission reaction. Concern over the potential for proliferation arising from the enrichment of uranium-235 was small, since it was believed that a nation, terrorist, or revolutionary group would find it easier and cheaper either to build a small reprocessing facility designed distinctly for the production of plutonium or to divert plutonium from a nuclear commercial reactor. Although this view was justified, given the huge costs and size of current enrichment (gaseous diffusion) facilities, recent developments suggest the launching of an era where the enrichment route to nuclear weapons proliferation from the commercial generation of nuclear power may become a reality.

In addition to gaseous diffusion-enrichment plants, there are also three other types of uranium-enrichment facilities, the commercial introduction of which will raise serious proliferation issues. Although the gaseous-diffusion process remains the dominant process used for uranium enrichment to date, the other processes are the gas centrifuge process, well proved technically and beginning to produce significant amounts of enriched uranium in Europe, two aerodynamic processes (South African and German) of which relatively little is known, and one unproved process based on newly acquired laser technology.

THE GEOPOLITICS OF NUCLEAR ENERGY

Although there is a critical difference of opinion concerning the adequacy of the world's uranium reserves to support a sizable expansion of the generation of commercial nuclear power, the availability of nuclear material (uranium, plutonium, thorium) may be influenced more by political factors than by an absolute scarcity of supply.[7] The breeder reactor, plutonium recycling, technological breakthroughs in enrichment technology (laser), and the development of the thorium-nuclear fuel cycle all offer opportunities for dramatically extending the growth of nuclear power and hence reducing the need for imported oil for the generation of commercial electric power.

Of these approaches, however, the development of the breeder reactor is qualitatively different in that breeders in principle increase fifty- to one-hundred-fold the energy obtained from uranium as compared to fission-generated light-water reactors, thus virtually eliminating scarcity as a constraint on energy supply for the twenty-first century. Although the doubling time of breeder reactors to date has not matched their theoretical capabilities, there is little reason to believe that these obstacles cannot be overcome.

A decision to develop the breeder could have a profound effect on the geopolitics of nuclear energy because it would reduce the need to mine or enrich uranium. As a result, countries possessing large-scale uranium reserves would see a decline in the demand for uranium ore, thus reducing the degree of their influence on supply. If a country possessing large uranium reserves, such as Australia, opted to wait for the development of laser-enrichment technology, which could prove more cost-effective than the breeder reactor, countries exporting breeder-reactor plants would compete intensely against those exporting laser-uranium enrichment plants.

If the United States, because of the availability of domestic energy alternatives, decides to follow the Carter program and slows its development of the breeder reactor, while its allies continue to develop the breeder, a situation could arise when the United States, having achieved a breakthrough in laser-enrichment technology, might find itself at loggerheads with its major allies in the international nuclear-power commercial market.[8] Likewise, if the economics or technological problems of laser enrichment hinder the development of this technology, then the United States could find itself at a competitive disadvantage vis-à-vis France and the Soviet Union and, to a lesser extent, Great Britain and Japan.[9]

These suppositions—for the technology is not yet available—have to be borne in mind and probabilities weighed if the worst aspects of competition between governments are to be avoided; the prospect is that they cannot be, which puts the highest premium on some form of multilateral management of a complex and awesome set of interests.

NUCLEAR ACCESS AND PROLIFERATION

The chief concern of the opponents of nuclear-power expansion is that the rapid expansion of commercial nuclear power is creating an ever-increasing amount of plutonium which may be diverted not only for use in the manufacture of nuclear weapons but also, owing to its extreme toxicity, poses grave world health and environmental dangers. Without major initiatives at both the national and the international levels, by the early 1980s about a dozen nations will have acquired fuel cycles which will give them the capacity to make nuclear weapons and will also make them vulnerable to the plutonium theft by terrorist, revolutionary, or criminal elements of their populations. More than any other factors, save possibly the general concern over safety of reactor design, fabrication, and operation, it is the risk of diversion of this dangerous fissile material that appears to be the pivotal link in how people feel about the dangers of nuclear power. Unlike thermal nuclear reactors, which require natural or slightly enriched uranium as fuel, the breeder reactor will initially use recycled plutonium contained in spent converter-reactor fuel. In order for this fuel to be recycled, however, it has first to be reprocessed.

If the economics allow, reprocessed fuel could also be used as new fuel for existing converter reactors. Thus, although the breeder reactor is dependent on the existence of reprocessing facilities, the danger of weapons proliferation, arising from the extraction of weapons-grade material from a reprocessing facility, could exist with or without the development of the breeder.[10] What the breeder does to the danger of weapons proliferation is to ensure that such reprocessing facilities will continue to proliferate.

From the best available evidence, the United Kingdom, France, West Germany, India, Japan, the Soviet Union, and the United States either possess or will soon possess commercial reprocessing facilities. (President Carter has opposed commercial reprocessing.) In addition, Argentina, Brazil, Iran, Italy, Pakistan, Spain, and Yugoslavia either have pilot-scale reprocessing facilities or have announced their intention to build their own commercial-scale reprocessing plants. Likewise, France has a quasi-commercial breeder reactor (Phoenix) in operation and the USSR has a much larger breeder facility in operation. Clearly, when nuclear reprocessing plants are owned and operated by a government, the potential for weapons proliferation is enhanced. This danger has to be weighed, however, against the tremendous rise in electrical-power generation that the breeder could provide to an energy-deficient world.

Although the United States has proposed that the plutonium problem could be controlled by having all nuclear reprocessing facilities concentrated in multinationally owned and operated reprocessing centers, this solution is likely to be politically unacceptable to many countries. For one example, the tough stance taken by the Carter administration on the 1975 German decision to sell Brazil a complete nuclear fuel cycle (including enrichment and reprocessing facilities) does not seem to have taken into account the Brazilian motivation in negotiating the agreement. The 1973-1974 oil crisis demonstrated to Brazil that it could no longer plan its economic development on the basis of a fuel whose availability was uncertain and whose escalation in price was devastating. Given the tremendous projected Brazilian growth rate for electrical consumption (especially in the São Paulo-Rio de Janeiro-Brasilia Triangle), the poor quality of Brazil's coal reserves, and the expansion of its hydroelectric capacity which is approaching its natural economic limits, Brazil must act to ensure her energy sufficiency. As a result, the Brazilians, who first expected to reach an agreement with the United States, consummated a nuclear deal with West Germany.

Although the Carter administration argues that its policy against Brazil's acquisition of nuclear enrichment and reprocessing facilities does not preclude Brazil's development of nuclear power but is designed to stop the potential proliferation of nuclear weapons, such a position goes to the heart of the question of access to energy. It raises the question of national sovereignty: To what degree can a nation allow a commodity so vital to its future to be controlled by others?

Brazil's overture to West Germany was prompted by the announcement by the United States on June 30, 1974, that it could no longer assure Brazil of the requisite amount of its enriched uranium needs. The Brazilians contracted for the relatively untested German aerodynamic enrichment process because this technology could be adapted to the thorium cycle once it becomes commercially viable. Brazil has some of the largest thorium reserves in the world and believes that the acquisition of the requisite technology will ensure its long-term energy independence.

Although there is no easy answer on how to assure access to commercial nuclear power without enhancing the danger of nuclear weapons proliferation, Brazil's cancellation of the Rio Pact, and its ability and announced willingness to find alternative partners

(South Africa, France, Soviet Union) if Germany were to accede to United States demands that Germany not meet its obligation to Brazil, warns again that questions of access to vital energy supplies—nuclear (or oil)—will involve the entire web of international political, economic, and military relationships.

THE POLITICAL COMPONENT

Although most studies on nuclear power focus either on the economic or weapons-proliferation aspects of an accelerated world development of commercial nuclear power, the political considerations that may play an important role in a country's decision to expand the nuclear sector of its energy economy have been neglected. The possession of a secure and abundant supply of electricity is of vital interest to every nation, and the decision to become self-sufficient in the raw materials and technology necessary to generate this power will be based on factors that transcend purely economic cost/benefit analysis. Once a nation turns to nuclear power to meet an ever-growing proportion of its energy needs, it is unlikely that it will settle for less than the possession of the complete fuel cycle and services necessary to operate such a program.

Although some states may believe their national sovereignties are not compromised by their dependence on others to provide some portion of these services (fuel fabrication, enrichment, reprocessing), there will be others whose genuine security fears (Pakistan), or national ambitions (Brazil, Iran), or international diplomatic isolation (South Africa, Argentina) make them insist on the complete possession of all the requisite ancillary facilities necessary for nuclear energy independence. Thus, ironically, a policy that seeks to curtail the spread of nuclear weapons by regulating the export of enrichment and/or reprocessing facilities may be the precipitating factor in another nation's decision to develop the breeder as a means of ensuring its access to vital energy supplies. Therefore, a central question will be: Should a country forgo the nuclear option if by possessing it its energy supply may be more assured? Alternatively, is energy security enhanced if the acquisition of a weapons capability sets off an arms race with one's neighbors who doubt the motivation behind the fuel cycle's acquisition?

THE BREEDER REACTOR

Although there is a security danger even with currently available converter reactors, there can be little doubt that the introduction of the breeder reactor, on a massive scale, will significantly increase national security concerns. These concerns will be exacerbated by the fact that, because the breeder reactor produces a very high concentration of the fissionable isotope plutonium-239, rather than the less-desirable plutonium-240 produced in converter reactors, weapons-grade material is produced in the normal commercial operation of the reactor, thus obviating the necessity for a costly additional weapons-producing facility.

Although the breeder reactor will escalate the qualitative and quantitative potential for nuclear weapons proliferation, the commercial development of gas centrifuge and laser-enrichment technology, as well as the development of national reprocessing facilities, will also underscore the security dilemma arising from nuclear power. It should be noted, however, that the failure to develop reprocessing and/or plutonium recycling facilities will not necessarily reduce the threat of weapons proliferation. It is also possible for a terrorist, criminal, or revolutionary group to gain access to the stored spent fuel and, with the requisite technological know-how, manufacture a crude nuclear-explosive device. Thus, unless these storage sites are well guarded, the nonreprocessing or recycling of spent fuel does little to alleviate the security problem. Likewise, the failure to reprocess the spent fuel before storage increases the danger of toxic substances escaping into the world environment.

URANIUM RESERVES AND RESOURCES

In order to assess the impact that the adequacy or inadequacy of uranium reserves may have on the expansion of worldwide nuclear power capacity, it is essential to examine the intricate forces which may determine whether uranium supplies are made available.

Uranium reserves are classified as "reasonably assured reserves" and "estimated additional resources." The grades of these reserves range from 0.05–0.20 percent of 10-40 pounds of U_3O_8 per ton of uranium ore. According to the OECD/IAEA (International Atomic Energy Agency) report on nuclear energy, reasonably assured reserves are estimated to be about one million tons of U_3O_8 of up to $15 per pound[11] of U_3O_8 and an additional 730,000 tons at prices of up to $30 per pound of U_3O_8. Estimated additional resources are estimated at one million tons of up to $15 per pound and 680,000 tons at prices of up to $30 per pound of U_3O_8.[12] Of this total, some 72 percent are believed to be located in Australia, Canada, South Africa/Namibia, and the United States.[13]

These estimates suggest no more than general orders of magnitude of volumes regarding worldwide available supplies of uranium. Reliance on them is hazardous as only a small fraction of the earth's crust has been adequately explored, the effects of recent price surges (1975-1977) above the $30 per pound level are not yet reflected in published reserve figures, and large, low-grade uranium reserves (not included in reserve statistics) may become economically attractive as the price of uranium ore rises. Because the cost of fuel is a small component of the nuclear fuel cycle, the price of uranium ore could rise rapidly before it would limit the expansion of nuclear power. For at least the next fifteen years, Australia, the United States, Canada, the Soviet Union, and South Africa/Namibia will continue to account for a major portion of both the world's uranium reserves and its production.

Unlike oil-exporting nations, most of these uranium producers are sizable consumers of energy. To the extent that these nations decide to go nuclear, this could have important ramifications on the availability of uranium ore elsewhere in the world. The fact that Australia, Namibia, and South Africa have reserves far exceeding their domestic requirements could serve to raise these nations substantially in the hierarchy of world economic power.

The demand for uranium ore will be determined not only by the demand for nuclear-generated electricity but also by the extent that plutonium reprocessing, new enrichment technology, and the implementation of the breeder reactor reduce the demand for uranium ore. Although several studies[14] have argued that the amount saved by these various technological breakthroughs will be volumetrically small (20-30 percent) and cost-ineffective, such savings may be of great interest to nations that lack both their own domestic uranium supply and enrichment facilities.

There can be no guarantee that even those states possessing vast uranium reserves will necessarily make them available to the world at large. Indeed, the nuclear debate in Australia on whether that nation should make its uranium ore available to the world, thus contributing to weapons proliferation, the curtailment by Canada of its nuclear exports in the wake of the Indian detonation of a nuclear device, the potential for a curtailment of South African and/or Namibian uranium ore, arising from internal or external political disruptions and transport problems in Gabon and Niger, are but present examples of situations which might affect the volume of uranium ore needed to fuel a major expansion in the world's nuclear-energy sectors.

Even if the availability of these reserves could be assured, the potential exists for a uranium ore-suppliers cartel.[15] When the prospects for the formation of such a cartel are viewed in concert with the Carter administration's efforts to place more stringent controls on the export of uranium enrichment, reprocessing, and breeder technology, the options for nonnuclear nations during the rest of the century become apparent;[16] namely, growing dependence on OPEC oil, total dependence on a uranium ore supplier and/or technology-oriented cartel, or acquiring some nuclear capacity (total fuel cycle) to lessen the degree of vulnerability to a cutoff of imported oil.

Ironically, it was the American decision in June 1974 to curtail the export of enriched uranium which largely precipitated the current uneasiness over reliable supply. The supply cutback to Europe and the notice to Brazil that the United States could not supply its commercial-import requirements led to both the Bonn-Brasilia nuclear deal and a rapid expansion in the breeder reactor and plutonium recycling programs, especially in Great Britain, France, Germany, and Japan. At the same time that the United States cut back its supplies of enriched uranium to foreign consumers it also urged the Europeans and Japanese in the International Energy Agency to accelerate the development of their nuclear-energy sectors in order to reduce their dependence on Middle Eastern oil supplies. A critical element of this diversification effort was the decision to develop commercial reprocessing facilities as a prelude toward implementing breeder-reactor programs.

Then, in October 1976, the United States shocked the international nuclear community by stating that it no longer viewed reprocessing as a necessary and inevitable step in the nuclear fuel cycle and that in the future the United States would engage in reprocessing and plutonium recycling only if they were found to be consistent with the nation's international objectives. The fact that the American moratorium on reprocessing and plutonium recycling made no mention of its breeder-reactor program startled the Europeans and Japanese and renewed fears that once again the United States was out to sabotage the growth of an independent nuclear industry in Europe and Japan. In the European and Japanese view, it appears that Moscow and Washington are alarmed about the possibility of being placed at a commercial disadvantage in the breeder-technology market, owing to the development of the breeder reactor by Europe and Japan. The deterioration in intraalliance nuclear cooperation was intensified by the introduction in the United States Congress of legislation designed to place further regulation on the transfer and use of American nuclear materials.

Although the Europeans and Japanese believed that the Carter administration might give them a more sympathetic hearing than did the previous administrations, this did not prove to be the case. The Carter administration called on the Europeans and Japanese to curtail their exports of those components of the nuclear fuel cycle that could lead to nuclear weapons proliferation (enrichment and reprocessing) and to forgo the development of the breeder reactor.

This last point is particularly distasteful to the Europeans and Japanese because, having embarked on a sizable research and development breeder program, they now find that they might soon be able to begin to market breeder reactors abroad. Some Europeans and Japanese are prepared to believe that President Carter called for a discontinuation of the breeder program because the United States was fearful that it might lose a large portion of its share of the world nuclear market to the others' technological advantage. The strong posture taken by the Carter administration against plutonium recycling and the development of the breeder reactor forms the crux of the ongoing international competition for access to the world's energy resources. This is particularly true concerning the Carter administration's opposition to third-world nations acquiring sensitive components of the nuclear fuel cycle. United States nuclear policy, as outlined by President Carter, seems to neglect the major question confronting the international nuclear community which is how the nuclear needs and aspirations of the nuclear "have not" nations may be met in developing nuclear energy for peaceful purposes.

While the International Nuclear Fuel Cycle Evaluation set up by President Carter in October 1977 is attempting to deal with the problems posed by the diffusion of sensitive nuclear technologies (heavy-water plants, enrichment, and reprocessing) by developing fuel cycles that are more proliferation-resistant, such technological "fixes," even if feasible, will take a minimum of ten years to develop and fifteen to twenty-five years to commercialize, thus taking us into the next century. In the interim, the rapid amassing of spent fuel necessitates vigorous action either in implementing a sizable spent-fuel storage program or in moving toward commercial reprocessing. What the United States must realize is that the world community is no longer willing to allow it alone to make the critical decisions that will affect the terms of access to the world's energy supplies, any more than the community of nations can accord such power to key oil exporters.

The nations of the third world realize that given the long lead times needed to develop alternative energy systems, 1985 is already history and by the end of the Carter administration's first term in 1980, 1990 may already be energy history as well. Decisions taken now will shape the energy future of generations to come; hence the necessity of selecting policies and making commitments of a long-term nature.

Because the nations of the third world are suspicious of the actions of the major nuclear-technology suppliers, particularly the United States, they are beginning to institute cooperative self-help programs in their nuclear development. Of particular interest are the relationships, however embryonic, that are developing between Argentina and Iran, Argentina and India, Iran and India, Argentina and Libya, Iran and South Africa, China and Pakistan, Australia and China, Brazil and Chile, and Argentina and Peru. There is also a burgeoning three-sided relationship among Germany, France, and China, and an unproved one between South Africa and Israel.[17] Although there is a tendency among nuclear policymakers to deprecate the significance of these relationships, they are important and should not be summarily dismissed. Ironically, to the degree that these initiatives are successful, the result of United States nuclear policy, namely, to curtail the potential for nuclear weapons proliferation, may be the reverse of what was intended.

THE NUCLEAR FUTURE

No sane observer of nuclear-energy development could recommend with complacency a continuation of present trends in the international nuclear system: 1) an increasing number of multilateral or bilateral deals for uranium ore and fuel-cycle technology, with political and other advantages to the signatories which are to the detriment of nonsignatories; 2) growing commercial and/or military conflict over access to uranium resources; 3) an accelerating waste-storage problem posing increasing health, safety, environmental, and security problems; 4) an almost certain rise in the risks of nuclear accidents and terrorism; 5) the potential for the further proliferation of nuclear weapons; and 6) a probable deepening of the divisions on nuclear fuel-cycle exports and breeder technology between Europe, Japan, and the United States on the one hand, and the nuclear-technology importing states and the third world nations on the other.

While the absolute importance of nuclear power in the world's total consumption of energy will remain relatively minor for the balance of the century, nuclear power will assume an increasingly important role in bridging the gap between the era of fossil fuel and that of renewable energy resources. With the world's oil and gas reserves falling and with the timely development of coal encumbered by environmental, technical, and economic obstacles, the rapid expansion of nuclear power is critical.

Although critics of nuclear energy would argue that we can forgo the nuclear option by such means as slowing economic development and introducing massive conservation, such arguments are unpersuasive and unacceptable to most nations. While no thoughtful observer would oppose reducing the fat in energy-use patterns, it is disturbing that those advocates who favor a return to a more pristine society, believing it to be more democratic than a system based on large centralized energy systems, seem unwilling to discuss what impact reduced levels of economic and energy growth will have on the economically marginal elements of the world's population.

The international challenge confronting us is stark. The extreme complexity of the issues necessitates the earliest possible consideration of an international process—comparable to what may also be created for oil in world trade—whereby the most rapid possible expansion of nuclear power can take place with multilaterally controlled and protected centers for enrichment, reprocessing, waste disposal, and the breeder reactor. Such a system cannot be hierarchical in nature; it must be based on the sovereign equality of all state actors in the international system. Our failure to encourage such a system could have the gravest of social consequences, especially, perhaps, for the industrialized world. The growth in new nuclear technology, laser enrichment, and fusion energy already threatens to outpace our efforts at regulation and control. If such oversight and control have never been achieved (or attempted?) on other matters affecting global war and peace, perhaps the time to begin is now. Because a decision to withhold the development of commercial nuclear power could have more ominous implications than would the decision to move forward, the timely resolution of the nuclear debate is imperative.

While no one believes that our long-term energy future will

not be bright, as hydrogen, solar, fusion power open up an era of energy abundance, such future opportunities must not divert us from one of the most fundamental challenges ever to confront mankind, namely, adopting mechanisms now to ensure that future generations pass into the era of energy abundance.

6

Ocean Frontiers of Energy

THE period of time from the present until the year 2000 has been described as marking the transition between energy generated by the nonrenewable sources—coal, oil, gas, and uranium—and energy created by plutonium or other fuel cycles, which extend almost limitlessly the sources of nuclear fission. Even if, for reasons of weapons proliferation or other technological advances, these cycles are largely forgone (perhaps to be replaced by other less-volatile fissile materials, such as thorium), we shall still be moving toward the time when access to energy will no longer be a problem for all societies.

A cardinal error committed too often is to assume that this era is upon us, that the technology exists, or nearly so, and that there is no need to undertake far-reaching energy policies and programs because the petroleum age is passing. Long lead times are required to bring additional conventional fuels on the market—a decade for a large and difficult oil field or a nuclear reactor based on existing technology. The thought is sobering: generally, we know now the contribution that civil nuclear energy will make to our societies by 1985; and we know how unlikely it is that the present importance of Middle East oil will have given way, by 1985 or even possibly by 2000, to new sources elsewhere. Thus, the perennial search for assured imported energy supplies will continue during the lifetime of most readers of this book.

Energy obtained directly from the sun is also a source that requires lengthy lead times. Even though the rudimentary applications of this source are commercially available, it will be several decades before the energy obtained from this source could reduce markedly our use of oil and our dependence upon foreign sources. Experimentation with the development of solar energy has spawned a number of related research projects which are only now moving from theory to laboratory (and to commercial applications, in a very few cases). Some of these test the use of energy derived from the oceans, energy that is limitless, renewable, clean, and of countless possible applications.

Exploitation of energy from the oceans will not be of any significance in our time, or during that of the next generation. Not until well into the third millennium will the oceans be tapped for the enormous energy which is contained therein. Energy from the ocean, like other sources of energy, originates with the sun, but the ocean is the largest collector by far. Energy from temperature differences in the oceans, winds, waves, tides and tidal currents, bioconversion, ocean currents, and salinity gradients are derived from the sun's effect on the earth's atmosphere. Ocean energy is attractive not only because it is renewable, in contrast to today's more conventional energy sources, but also because it appears to be an environmentally acceptable source. Until recently, however, the technology to develop these energy sources was too expensive to be considered practical. The ocean as an energy source may have its first use in limited areas where, for reasons of current, tides, and winds, the generating plants are close to the consumer.

Experimental units of power plants that exploit ocean waves, wind currents, and the temperature difference between the surface and bottom layer of the ocean will be in operation by 1985. It will take until the year 2000, however, for energy to be produced in commercial quantities from the ocean and far longer for it to be of great consequence in world energy consumption (with the exception of particular localities). In the meantime, legal and political arrangements are already being negotiated which can provide a proper context in which these new and varied energy endeavors may proceed.

INTERNATIONAL CONSIDERATIONS

Interest in developing the ocean's energy potential does exist. Even if the necessary technology is developed in an efficient manner, however, the concerned states must also deal with the complicated jurisdictional matters involved with the international exploitation of the oceans. The development of these potential reserves will be connected to whatever international legal guidelines are established, since many of the prime locations for anticipated systems lie outside territorial waters (where national sovereignty is virtually complete) and also in areas where two or more states are seeking to extend their control.

Future regulations concerning the use of the ocean's energy potential need to be established, perhaps in the current negotiations for a Law of the Sea and also in the actions of the states. Although these are new uses of the oceans which were not anticipated in 1958, when the Conventions of the Law of the Sea were drafted, Article 2 of the Convention on the High Seas states: "freedoms . . . which are recognized by the general principles of international law, shall be exercised by all States with reasonable regard to the interests of other States in their exercise of the freedom of the high seas," thus providing for the states' use of the high seas for other, nonstipulated purposes, of which energy development is one.[1]

The territorial sea, in the past, has been established at three miles from shore, beyond which lay what is considered the high seas. Many states, however, have been independently extending their zones of jurisdiction. There is agreement among those attending the Third United Nations Conference on the Law of the Sea to the extension of the territorial limit of coastal states to twelve nautical miles and, additionally, to give them the right to exploit resources within an "Exclusive Economic Zone" now anticipated to extend seaward some 200 miles. The Exclusive Economic Zone, as presently drafted, would give to the coastal state "sovereign rights . . . for the economic exploitation and exploration of the zone, such as the production of energy from the water, currents, and winds."[2] Even if the current Law of the Sea Conference fails to approve this extension of jurisdiction over resources, there is every reason to believe that most states will, nevertheless, summarily seize possession of the region covered by the proposed extensions. In both cases, therefore, it appears that the area designated the high seas will be reduced, thus setting the stage for future conflicts over the exploitation of the ocean's resources.

Except for salinity gradients and tidal plants, most large ocean-derived energy facilities will be installed outside the territorial sea; therefore, they will be outside the zone of the coastal states' general sovereignty. Although the first offshore geothermal-energy power plant is likely to be on the continental shelf, this would place it within a state's jurisdiction under the articles of the Convention of the Continental Shelf, not within its area of sovereignty.[3]

It will not be until far into the next century that it will be technically, economically, and politically possible to exploit geothermal deposits beneath the deep ocean floor, i.e., below the high seas. At that point, these resources might fall within the power of an international regime, established and governed by the United Nations. The experience of states with the proposed international regime in the mining of manganese nodules will have an impact on how the geothermal resource may be tapped.

Although the exploitation of waves, wind, and other energy sources from the water column would not fall under the Convention of the Continental Shelf, which gives to coastal states the exclusive right for resources of the continental shelf proper, these potential forms of energy could possibly be exploited under two other agreements. The first would be through the proposition of the Exclusive Economic Zone, currently being negotiated. The second could come under the principle of "reasonable use" of the high seas, to which all states have the right, as long as such uses do not interfere with other states' "exercise of the freedom of the high seas."[4] The inclusion of the right to exploit the ocean's energy potential specifically within the negotiating text would be a far-sighted provision, anticipating the eventual development of this untapped resource.

While United Nations Resolution 2749 declares that only the resources of the deep seabed are the "common heritage of mankind," beyond the limits of national jurisdiction, all states would have an equal right to exploit resources, under the "reasonable use" doctrine. It is likely, however, that the developing countries will attempt to place any exploitation beyond national jurisdiction under an international regime, so that these states might have a greater share in the profits than they would receive if the development of the energy resources of the high seas was privately controlled. If this is the case, the exploitation of the ocean's resources, including sources of energy, would be delayed even further.

There are a variety of possible conflicts which can arise from the establishment of energy-development plants situated outside the 200-mile limit. Pollution can be one source of major conflict. While the Stockholm Conference declared a nation responsible for environmental damage caused by activities under its jurisdiction, it does not make allowances for the various pollution standards of different nations. On the high seas, a plant (which would most likely be owned by a multinational corporation[5]) could fly the flag of any nation. A flag of convenience might guarantee more profit by allowing the plant to operate under lower pollution standards than those of an adjacent coastal state.[6]

Perhaps various regulations which are or will be in effect will help alleviate any discrepancies in pollution standards, by applying guidelines for all areas. The 1972 Ocean Dumping Convention[7] regulates the deliberate disposal of waste from vessels, platforms, or other man-made structures. Another international agreement that would control activities on installations of the kind involved in ocean-energy exploitation is the 1973 Intergovernmental Maritime Consultative Organization Convention for the Prevention of Pollution from Ships.[8] A further area that might be affected by pollution from offshore installations—kelp farming and the harvesting of marine cultures—might be regulated by the Food and Agriculture Organization of the United Nations.

Energy development in the high seas could also possibly conflict with the activities and interests of the adjacent coastal state. Offshore installations could interfere with or exclude various activities of the coastal state, such as the fishing and shipping industries, since these installations are clearly navigational hazards which can hinder the safe passage of ships both within the Exclusive Economic Zone and on the high seas.

Although the Convention on the Continental Shelf established a 500-meter safety zone around installations located on the continental shelf, whose purpose is the exploitation of resources of the continental shelf, there are no provisions for plants that exploit the water column. The Ocean Thermal Energy Conversion (OTEC) installations are very much a case in point. Because of the requirements of cold water pipe, these facilities are moored in great depths, requiring long anchor chains. Consequently, they can swing over large areas, thus potentially interfering with the navigational rights of other nations. With ocean space becoming more crowded, the coastal state might have to assign safety fairways (as in the approach to deep water ports) to permit ships to navigate around such obstacles, thus further extending controls over the transit of these waters.

Another source of contention is the possible limitation to naval operations that such structures may impose, as well as the use of offshore installations for intelligence reasons. Structures such as OTEC plants may interfere with the routes of naval ships; depending on location, the mooring cables and cold water pipes could interfere significantly with the passage of submarines. Conversely, while detection devices can be planted on the installations themselves, the possibility of submarines hiding near OTEC plants to avoid sonar detection also exists. Further, these platforms could conceivably be used to collect intelligence, an activity that has long generated deep mistrust on the part of the coastal states. These countries have been especially wary of the uses to which the superpowers may put such facilities, thereby infringing on their neutrality.[9]

Despite such possible conflicts, energy derived from ocean sources offers many countries the prospect of an alternative energy source. Present technology for most thermal-energy plants indicates they may be located in the southern hemisphere, off the coast of developing countries. For these states, there is the chance of bypassing some of the period of oil and gas through the exploitation of energy sources off their coasts.[10]

Nevertheless, the development of the necessary technology to exploit the sources of energy from the ocean requires an enormous amount of time and money. Yet, even though the day is far-off when these installations can be operational, a framework of international law is already developing. In terms of access to energy, this is one area in which an established law will help limit the potentially corrosive political effects certain to be the consequences of competition between governments in the acquisition of choice locations for the development of the ocean's energy resources.

ENERGY SOURCES FROM THE OCEAN

In highlighting the following areas of research and technology in ocean-derived energy, I am indicating only some of the more promising undertakings that illustrate the variety of means available for extracting energy created by the solar effects upon the oceans.[11] The development of these sources, in some cases, will be of greater benefit to some of the developing countries, whose commercial-power requirements are of a lower-order magnitude or where special coastal configurations allow for such innovative enterprises.

Ocean Thermal Energy. One of the more exotic, yet most promising, sources comes from the stratification of ocean water. Surface water collects the energy from the sun, while water at 1,000 to 4,000 feet in depth is about 4.5° C colder, due largely to currents from the poles. Ocean Thermal Energy Conversion technology uses this temperature difference between the surface and the bottom layer.

The technology involves a closed-cycle system. A working fluid with high-vapor pressure (ammonia, freon, or propane) is circulated through an evaporator and heated by warm surface water. This changes the working medium from liquid to high-pressure gas. The vapor passes through a turbine generator which produces electric power either for transmission or chemical processes or, conceivably, for an energy-intensive industry. The vapor is condensed by cold seawater from the ocean's depths and this working fluid is then pumped back through the evaporator.

A sea thermal-power plant can be located offshore on a platform, with the cold water pipe reaching 1,000 to 4,000 feet deep, the platform being anchored to the ocean floor.

Possible sites for OTEC plants include the Gulf of Mexico, the Gulf Stream off Florida, Hawaii, and the Caribbean. Since large temperature differences are favored, the best locations are between the Tropics of Cancer and Capricorn. The selection of the appropriate sites would be based on the existing winds, waves, currents, and degree and stability of temperature difference.

For the United States, a power plant in the Gulf Stream off Florida would be able to deliver electricity via cable or energy in the form of high-pressure hydrogen. In the Gulf Stream, where water flows rapidly, there would be less of a problem of temperature layers breaking down. In such locations, however, OTEC plants would have to be built to withstand stronger currents and winds than those in the tropics, and they would operate with a much lower temperature difference. These plants could be built off the continental-shelf break where larger temperature differences can be obtained in shallower water.

Offshore thermal-energy conversion is attractive because it allows energy to be produced on a continuous, unbroken basis with minimal environmental impact, is based upon a renewable source, and could be used by utilities as a baseload if not always as the sole source.

Ocean Waves. Waves may be used in three ways to exploit their energy potential. The first method involves the rise and fall of following waves to build pressure which then drives a turbine.

A second application is illustrated by a promising and novel laboratory design which exploits the rolling motion of waves. In this proposal, vanes or cams are aligned on an axis and then submerged to a depth of about twenty meters, with only one meter showing out of the water.[12] The movement of the waves against the vanes produces pulses of high water pressure which in turn drives a turbine located at the center of the structure. One power-generating unit would be composed of twenty to forty segments of a total length of 500 to 1,000 meters. This concrete floating structure is then towed out into the ocean and drifts under the impact of the waves. A problem affecting such designs, as it does virtually all present, seemingly visionary, processes for obtaining energy from the oceans, lies in the extremely difficult and costly long-distance transmission of power. There may be no greater challenge in energy research and technology than the one of "distant" power. A solution may be a process whereby wind-wave temperature-gradient sources of energy are used to produce hydrogen from the oceans, transmitting it as fuel by ship or pipeline. Hydrogen can then be used ashore either as gas or in electric-power generation.

A third application of wave energy uses the power that would be gained by directing waves into narrow and flat channels until the head or pressure of the water is strong enough to drive a turbine.

While the potential for wave-generated power is enormous, waves will not provide continuous energy. Therefore, an alternative source will have to be available. In the United States, for example, the best region to use wave power is off the coasts of Washington and Oregon, but only during the fall and winter months. In the Atlantic off Great Britain, wave energy could be produced 80 to 90 percent of the time; nevertheless, it would still require an alternative source.

Wave power is attractive as a permanent and clean energy source. As always, however, there are environmental consequences that have to be considered. Beach water might become cooler because of the energy absorbed by the station. Disturbances may result in the erosion of beaches and the silting of harbors. Consequently the location of the wave-power installation has to be considered carefully. If it is assumed that these related effects are controllable or at least acceptable, the key difficulties in water-power exploitation may lie in the transmission problem and in the lapse periods when wave action is insufficient.

Ocean Currents. Another potential source of energy is the harnessing of the ocean currents. The concept for tapping them is simple. A series of waterwheels set up on the floor, each with twelve blades measuring 100 meters in diameter, has been proposed. In this method, they would be located in the Gulf Stream as it passes through the Straits of Florida. Other designs call for different types of rotors, propellers, and turbines; one proposal advocates parachutes be attached to a continuous rotating belt. In this case, the parachutes would be pulled with the current and collapse when they move against it.

Some of the obstacles involved with the use of energy derived from the ocean currents are that there are few regions where the currents are strong enough to make the extraction of energy possible (potential sites would be located at the Straits of Gibraltar, Bab el Mandeb at the mouth of the Red Sea,[13] and the equatorial Cromwell undercurrent [1.5 miles west of the Galapagos Island]), and it is difficult to transmit energy over long distances, once it has been extracted.

Tides. Energy plants operating on the continuous movement of the water of incoming and outgoing tides drive a turbine for power generation. In order to produce electricity continuously, the blades in the turbine have to be adjustable to different pressures of water through the tidal cycle and to be reversible when the tide turns.

Such an energy source is being exploited in France, across the estuary of the River Rance, and has long been a dream for the Bay of Fundy in Nova Scotia. Nevertheless, the application of tidal energy is limited, since an average tidal range of at least sixteen feet is presently required to make it economically feasible. In the United States this condition is found only in the Bay of Passamaquoddy (part of the Bay of Fundy) and Cook Inlet in the Gulf of Alaska. Locally, however, these tidal power plants could become important as totally renewable sources of energy.

Another tidal plant is operating in the Soviet Union, between the White Sea and the Ura Guba Bay. There are further prospective sites worldwide, among them, in the English Channel and the North Sea, off British Columbia, off the coast of Korea, off China, in the Gulf of Bengal, off Pakistan, off Western Australia, and in southern Argentina.[14]

The basic technology exists and will certainly be improved. Tidal power provides clean and renewable energy; its environmental impact rests on the movement of silt.

Wind. Wind power will also be contributing to an energy self-sufficient society. The greatest potential for the United States is along 1,000 miles of the continental shelf off the northeastern coast. There the wind blows in a strong, prevailing pattern, more powerful than over land as speed appears to intensify with distance away from shore.

Thirty-five-foot-diameter wind "wheels," 200-620 feet above sea level, would turn generators mounted on concrete floating platforms, anchored to the ocean floor, and connected to an electrolyzer platform which will produce hydrogen. Wind-power stations, in order to be effective, must be built far offshore and, based on present technology, would thus be unable to transmit electricity back to land.

An alternative for this scheme—and for many other ocean energy-gathering stations—would be to use the energy provided to manufacture energy-intensive products such as ammonia for agriculture and chemical needs, or aluminum from alumina.

Another drawback is that wind energy is not dependable: wind blows with different strengths at different times. At times of low winds, an alternative energy source or stored energy must be available. Wind generators as presently designed may not be able to withstand great storms. These plans are still largely in the experimental stage.

Nevertheless, many sites will be suitable to generate wind power such as the continental shelf off Nova Scotia, west of Ireland, on the eastern edges of the North Sea, the shelf off South Africa and Australia, and off the islands in the trade winds and the roaring "forties."[15]

Oceanic Bioconversion. This source of energy is derived from a combination of nutrient-rich ocean water and the sun to create varieties of seaweed and algae which can then be processed to yield methanol and other fuels. As an example, one variety of kelp grows as much as three meters a day in waters of about 20° C and has been selected for commercial farming at an experimental site sixty miles west of San Diego, California. The kelp requires cold, nutrient-rich water to grow, either at the site of a natural upwelling or one from mechanical pumping devices. (Current experiments involving kelp use pumps driven by wave action; future plans include the possibility of using deep water discharged from offshore thermal-energy conversion plants.) A portion of the seaweed will be harvested about four times per year, then dried and processed. It has been estimated that a commercial 100,000-acre farm could be in operation by 1985, providing an annual yield of 100,000 kilowatt hours per acre from processed seaweed tissue.[16]

Geographical areas favorable to kelp farming are the cold and moderate-temperature regions of the Atlantic and Pacific oceans that are close to shore, especially in areas where the natural upwelling occurs. While this particular kelp will not grow in tropical waters, there are alternative strains which could be harvested.

Oceanic bioconversion could have adverse effects upon the surrounding environment, owing to the changed temperature of the water. The installations necessary to extract energy could also interfere with established shipping routes.

Ocean Geothermal Energy. In contrast to other kinds of energy that can be extracted from the ocean, ocean geothermal energy is nonrenewable, although the implications of this distinction, important in the case of oil, are of minimal consequences. Geothermal energy is the heat of the earth itself contained in hot water and rocks situated deep within the planet. It is currently being used onshore in locations where the deposits are relatively close to the surface. In these cases, it is used to generate electricity, as well as to provide heating for residential and industrial areas.[17] While geothermal deposits have been located offshore, the current technology is not yet enough advanced to develop it effectively.

Nevertheless, ocean geothermal reserves are an almost limitless source of heat, especially if they can be exploited in such locations as the Mid-Atlantic Ridge, the East Pacific Rise, and the Red Sea.[18] Great Britain has plans to explore the geopressured energy beneath the North Sea once the oil and gas reserves have been depleted, using the existing platforms. And within the United States there is an extension of onshore geothermal energy in the Gulf of Mexico.

Despite these known locations, there is still uncertainty over the technology of developing the geothermal energy potential in the deep ocean. Therefore, there appears to be little probability of commercial exploitation before the middle of the next century.

These varied devices by no means exhaust the range of possibilities that the ocean contains. They should be regarded as illustrative only of some of the types of systems of which we shall read more as we move through the petroleum fuel era well into the next century.

We cannot assume too readily that access to energy from other sources, especially the oceans, will offer fewer problems than has access to oil or nuclear power. But if the history of oil tells us that efforts toward accommodating differences between producers and consumers now require institutional arrangements, and that nuclear issues as they arise are being dealt with (we hope) by innovative cooperative proposals for minimizing risks, we see in the energy role which the oceans can play a prior arranging of international laws and institutions aimed at resolving issues between states, before they arise, as it were.

Conclusion

FROM the outset of this inquiry into access to energy, my basic assumption is that until well into the next century the great energy-consuming nations of the world will continue to depend upon the importation of energy, excepting perhaps only the Soviet Union. Most countries simply do not have adequate indigenous supplies of oil, gas, and uranium ore. For a favored few, such as the United States, Canada, and the Soviet Union, coal in its direct fuel state or in modified form may prove to be (once more) an exceptionally bountiful resource. But for other fossil fuels, inadequate domestic reserves in the longer term is thought to be the more likely and general situation. It is only prudent to act as if it were to be the case, for no way is known of measuring what resources we have yet to discover.

There are two hopes that, if they materialize, might prolong by several decades the contemporary petroleum era: first, substantial improvements in recovery techniques—the ability to obtain a greater percentage of the oil in existing fields than is now (30 percent) the general case. An increase in that rate of only 1 percent might add twenty-five billion barrels to the world's inventory of recoverable oil. Second, discoveries of vast reserves, comparable to those of the Middle East. Judgments vary as to the likelihood of such good fortune. Some assert that it is a strong possibility, but most analysts clearly warn that we have no sufficient justification, on the basis of experience, for expecting such discoveries to be made, or to be made in time to forestall competition among states over access to existing reserves. The more optimistic point to Mexico and the higher estimates of its oil asset as an example refuting those who hold our prospects are not likely to be bright. But no one yet has a truly considered judgment on the scale of the Mexican discoveries. Such cannot be obtained until very much more is known. Even when we think the size of the reserves is known, technical, economic, and political factors will shape Mexico's decision as to how much oil should be produced, how much kept for domestic use, and how much sold into world oil trade and at what price. Only one point is certain: the discoveries are good news for Mexico. If the more prudent estimate of proven reserves of forty billion barrels proves to be more nearly correct than estimates of several hundred billion barrels, Mexico is not likely to add more than three million barrels a day to the volume of oil in international supply—not a major addition—perhaps 8 percent. Thus the warning signal that oil-importing states will be contending with each other for available volumes is not cancelled by any reasonable forecasts of additions to supply from new discoveries. Nothing in sight suggests we can delay intensive research into other sources of energy.

Significant advances in energy technology could materially alter the conventional energy position of many states by drawing upon geothermal, fusion, and solar sources—especially the last, which promises eventually to bring to mankind unlimited quantities of energy. Will the less developed or industrializing states take earlier and greater advantage of new techniques? They are searching for ways of bypassing the age of petroleum and, partly because of their much lower rate of energy consumption, may be able to do so.

Nevertheless, for the principal energy consumers, questions of access to sources of energy could continue to dominate energy diplomacy into the next century, almost as it does today. The very long lead times that lapse between even great scientific discoveries and their introduction into the marketplace have usually been estimated at twenty-five to fifty years; for a great scientific advance to alter fundamentally the economic base of an advanced society it might take much longer.

It can be argued then that access to energy will be important and possibly still crucial. By "access" we have meant the terms on which one society is able to obtain sufficient, additional resources to supplement its indigenous reserves. There may be no older problem than this: a people needing the materials possessed by another.

Traditionally, such needs have been met by raids or larger military actions leading to physical occupation of another's land with eventual political control implanted over the alien society. If the commodity is less than vital, trading would be the means; but when a material, for any reason, becomes crucial, then, to our lasting regret, attempted seizure is the rule. Throughout history, force—its threat or use—has been the common ingredient in ensuring one's access to the resources of another.

With regard to energy, and petroleum in particular, the record of means of access includes physical occupation (Japan seized Indonesia in 1941); a "presence" of an imperial power which once assured supremacy (Great Britain, Kuwait, and the United Arab Emirates); and domination of a country's oil by the commercial interests of another state such as the virtual monopoly over the disposition of Venezuelan oil by the United States (particularly) and Great Britain. There were instances in which a major power's objectives involved a reach in part for the oil of another: Germany and the Caucasus; the Soviet Union and Iran.

The colonial-type experiences of all major exporters of oil have deeply colored and permanently affected the terms under which oil is now and will be obtained. If we define "colonial-type" as embracing a military action to secure or preserve access to someone else's oil, or occupation, political control, and/or exclusive power to dispose of the resource in volumes and at prices set by the metropolitan power, then we list France, Japan, Great Britain, and the United States; and we number among the "colonial" lands not only Indonesia, Mexico, Venezuela but also Saudi Arabia, Iran, Iraq, Syria, and even Canada and Australia, not usually found in such company. But in these two nations, the predominant role of foreign capital and foreign control of their oil has been widely considered to have resulted in a history of outsider's control over the resource in question.

For nearly a half-century, oil in the world was controlled and obtained by another nation largely through the concession system. Varying in details as well as in scope, the "concession" came to represent in odious form foreign domination. These arrangements could not have survived long in the post-World War II environment, which made plain both the inability and unwillingness of western empires to defend what they had regarded as their own. It was inevitable, perhaps, that the emerging economic and political nationalism of "new" states should come to be centered on the concession system, which had become symbolic both of exploitation and a vital diminishment of the host government's sovereignty.

This was fully and clearly understood by Mohammed Mossadeq, who, with nationalistic fervor, caused the concession arrangements in Iran to be altered profoundly. Perhaps particular credit must be given to General Ibnu of Indonesia, who acted upon the concessions system in a less flamboyant but equally pervasive way. General Ibnu insisted that as long as Indonesian oil was in the hands of foreign oil companies, the domestic political pressures this provoked could become uncontainable. His point was basic and perceptive: change the concession system to one of contract with the Indonesian government and do so in a timely fashion before political pressures become acute.

As seemed to happen everywhere, the oil companies with entrenched and large interests fought such concepts and policies until compelled to give in, more often than not because smaller companies sought entry and were willing to experiment with new forms. In this period (1956-1970), now lacking the support even of home governments in the West, the companies gave in; profound changes in access resulted.

The OPEC revolution capped the collapse of the oil companies' familiar system of total control by initially demanding that it be a government responsibility to determine the terms for oil in world trade. OPEC then insisted upon participation or part ownership in an oil company's venture, then majority share and ultimately, in many situations, a takeover which, in effect, left the companies working really in behalf of the producer government but still providing the complex logistical expertise essential to the worldwide movement of oil.

These events could not have occurred in isolation from more general developments in the international political and economic system. The whole fabric of the colonial system for assuring access to raw materials, not just oil alone, was being torn as the "winds of change" swept the former colonial world. The victory of nationalism in challenging the system created by the West was considerable; what is still not evident is the nature of the system that will succeed it.

The earlier arrangements were one-sided. While the imperial/colonial system resulted in some undoubted benefits to the host countries, no one would contest that the scales were tipped heavily in favor of the oil companies. It is with the greatest reluctance, therefore, that some international companies have come to understand that their critical interest lies not with any particular form or mode but with *access*, almost regardless of how the contractual relationship with government might be expressed; so it was with oil and now with uranium.

There is nothing unique or inherently unreasonable in the desire of producers/exporters to improve upon the terms of trade or to withhold supply against future need and greater value, however difficult the issues may be. Nevertheless, the contemporary world is slowly becoming aware of the profound interests all parties have in creating processes that advance the mutual dependence of economies and societies.

Recent events in Iran serve only to emphasize these points. Not only did oil (this time, the revenues obtained from its sale) become again a profoundly disturbing factor in Iran's national affairs, but the disruption that accompanied and followed upon the collapse of the Shah's regime led to great uncertainty as to the adequacy and continuity of oil in world trade. Not only were questions asked about the security of the Persian Gulf in general, and the durability of the Saudi kingdom in the face of somewhat comparable internal pressures pitting the old against the new, but also questions were asked about the willingness and ability of the United States to remain the principal external power able to influence events.

The very anticipation—and then the reality—of a loss of Iranian oil exports precipitated the most serious supply issue since the winter of 1973-1974. Until Iran achieves a stable national government, access to its oil will be problematical. The question is not only technical—how much can be produced—but political as well: what level of production will satisfy the economic objectives of the government? meet the defense requirements? The decision may be to reduce exports or to attain the levels reached in the Shah's era. We cannot yet tell, but Iran is a key example of how greatly changed are the considerations that lead to decisions on oil production. Should Iran choose a lower level than will meet the increasing demand of industrial and developing states and should other producers not make up the difference, then the issue of the right of any government to withhold supply of a commodity vital to the international community comes front and center.

Perhaps the most perplexing challenge inherent in the dialogue between the industrial world and developing nations is this question of sovereignty over natural resources in the context of an increasingly interdependent world. Even raising the issue today runs the risk of embroiling all possessors of raw materials, and the industrial world, in a maelstrom of charges of imperialism.

In oil, we can look to a continuation of the present situation in which oil in world trade—in the volumes required by the importing states—will come not only largely from the developing world but from the Persian Gulf. And in that region, the decisions of Saudi Arabia and Iran about the amount and price of oil exports can make the difference between war and peace. Less than a decade from now, Iraq might have near comparable power. Since it has won so recently the right, as a sovereign state, to dispose of its resources almost at will, what is now to be said concerning the "right" of a society such as Saudi Arabia to withhold supply of a commodity that has become vital to others? Curiously, it is the Soviet Union that has raised this question, when a Soviet source referred to oil in the Middle East as "international property."

The question of a society's "rights" must also be asked about commodities other than oil: of food—and the role of the North American continent as breadbasket to the importing world—and of uranium ore, and the role of South Africa, Canada, Australia, and the United States. It cannot be posed in the context of energy alone, although energy and food have the highest priority.

It would seem evident that there can be no untrammeled "right of sovereignty" in these regards; in the international arena, the right comes to be circumscribed by customary law and treaty.

Some form of international device or process is needed to deal with the complex issues of "need," "value," "development," and "access" to the world's high-priority commodities, a process that must involve equally both producing nations and importing nations. Given the imperfect record of the human race in such situations, the impetus for such an understanding is more likely to come in the aftermath of a shock occasioned by a producer's withholding supply "unreasonably" and a consumer's believing himself faced with no alternative but to act forcibly.

It is the central issue, nevertheless, posed by interdependence. If the institutional process can be created for oil and uranium ore, and with the nuclear fuel enrichment cycle, its example might be contagious. Instead, we have been watching something that has a risk of becoming a paramonopoly of nuclear supplies capable of holding other states at some lower nuclear level.

As the greatest single consumer and the world's largest producer of energy, the United States will have to play a leading role in fashioning such an arrangement. Yet, this would not be an exclusive role. At this point we can see the significance of such undertakings as the Common Market's Euro-Arab dialogue and the Lomé Convention, an experiment at achieving a more satisfactory and durable understanding between those countries in the Common Market and forty-six states of the former colonial world. Such undertakings provide the basis for the fair and peaceful distribution of the essential commodities, fair both to those who own the resources and to those who need those resources. The essence of these efforts is that the inherent differences of interest in access to energy commodities be resolved before they spin out of control; the object: peace in our time, and for generations to come.

Notes

Introduction

1. *World Energy Outlook*, Report of the Organization for Economic Cooperation and Development (Paris: OECD, 1977).
2. *The International Energy Situation: Outlook to 1985* (Washington, D.C.: Central Intelligence Agency, April 1977).
3. A new generic term for energy sources emanating from the sun: solar power, wind, ocean currents.

Chapter 1

1. Herman T. Franssen, *Towards Project Interdependence: Energy in the Coming Decade*, Prepared by the Congressional Research Service, Library of Congress, for the Joint Committee on Atomic Energy, U.S. Congress (Washington, D.C.: Government Printing Office, December 1975).
2. Jeremy Russell, *Energy as a Factor in Soviet Foreign Policy*, Published for the Royal Institute for International Affairs, London (Lexington, Mass.: Lexington Books, 1976).
3. *Project Interdependence: U.S. and World Energy Outlook through 1990*, Prepared by the Congressional Research Service, Library of Congress, for the Senate Committees on Energy and Natural Resources and Commerce, Science, and Transportation, and the House Committee on Interstate and Foreign Commerce (Washington, D.C.: Government Printing Office, June 1977), p. 60, and *The International Energy Situation: Outlook to 1985.*
4. Amory Lovins, "Energy Strategy: The Road Not Taken?" *Foreign Affairs* 55, no. 1 (October 1976): 65-96.
5. Carroll Wilson of the Massachusetts Institute of Technology, organizer and director of the Workshop on Alternative Energy Strategies study (New York: McGraw-Hill, 1977), believes coal may well be of far greater consequence in world energy trade if the requisite effort to develop coal reserves is undertaken.
6. Mason Willrich and Melvin A. Conant, "The International Energy Agency: An Interpretation and Assessment," *American Journal of International Law* 71, no. 2 (April 1977).
7. See Charles K. Ebinger, "Resource Conflict and Conflict Potential in the Southern Hemisphere: A Preliminary Assessment," paper prepared for the Institute for Foreign Policy Analysis, Washington, D.C., April 1977.
8. For an analysis of the potential supply/demand gap, see *The International En-*

ergy Situation; World Energy Outlook; Walter J. Levy, "U.S. Energy Policy in a World Context," *Petroleum Intelligence Weekly* (April 11, 1977); and *Project Interdependence.*

9. Geoffrey Chandler, "The Next Energy Crisis," Address to the Manchester Statistical Society, November 9, 1976, p. 1.

10. Franssen, *Project Interdependence,* p. 58.

11. Chandler, "The Next Energy Crisis."

12. Ibid.

13. Putnam M. Ebinger, *Division and Realignment: Chinese Energy Diplomacy in Europe and the Middle East,* Monograph Series (Great Falls, Va.: Melvin A. Conant/International Energy, June 1977).

14. *World Energy Outlook,* p. 57.

Chapter 2

1. Neil H. Jacoby, *Multinational Oil* (New York: Macmillan, 1974), p. 30; and John M. Blair, *The Control of Oil* (New York: Pantheon Books, 1976), pp. 54-56.

2. Jacoby, *Multinational Oil,* p. 34.

3. Ibid., p. 58.

4. *The Relationship of Oil Companies and Foreign Governments;* Report prepared by the Office of International Energy Affairs, Federal Energy Administration (Washington, D.C.: Government Printing Office, June 1975), p. 121.

5. Franklin Tugwell, *The Politics of Oil in Venezuela* (Stanford, Calif.: Stanford University Press, 1975), p. 39. His study, and other writings, have contributed greatly to our understanding of the history of international oil.

6. Zuhayr Mikdashi, *A Financial Analysis of Middle East Oil Concessions: 1901-1965* (New York: Frederick A. Praeger, 1966), p. 155.

7. Blair, *The Control of Oil,* p. 79.

8. Mikdashi, *A Financial Analysis,* p. 157.

9. Blair, *The Control of Oil,* pp. 80-90.

10. U.S. Congress, Senate, Committee on Foreign Relations, Subcommittee on Multinational Corporations, *Iraq Petroleum Situation: Hearing on Multinational Petroleum Corporations and Foreign Policy* (State Department Memorandum for the Under Secretary from Andreas Lowenfeld), 93d Cong., 2d sess., October 24, 1964, pp. 8, 491.

11. Blair, *The Control of Oil,* pp. 86-90.

12. Ibid., p. 89.

Chapter 3

1. Geoffrey Barradough, "Wealth and Power: The Politics of Food and Oil," *New York Review of Books* (August 7, 1975), p. 23.

2. Robert O. Keohane, "International Organization and the Crisis of Interdependence," *International Organization* 29, no. 2 (Spring 1975): 357 ff.

3. "Bilateral Deals: Everybody's Doing It," *Middle East Economic Survey* (January 18, 1974), p. 1.

4. Willrich and Conant, "The International Energy Agency."

5. For a detailed assessment of the IEA, see ibid., pp. 200-223.

6. Roger D. Hansen, "The Crisis of Interdependence: Where Do We Go from Here?" *The United States and World Development: Agenda for Action 1976* (New York: Praeger, 1976), p. 48.

7. For an elaboration of the LDC demands, see "Declaration and Program of Ac-

tion concerning the Establishment of a New International Economic Order," United Nations Sixth Special Session.

8. See "Overall Proposal for Measures in Favor of Development and International Cooperation," Statement given by Oil, Finance, and Foreign Ministers of OPEC, Algiers, 1975.

9. Speech by Secretary of State Henry A. Kissinger before the National Press Club, Washington, D.C., February 3, 1975.

10. Ibid.

11. Ibid.

12. Barradough, "Wealth and Power," p. 29.

13. Hansen, "The Crisis of Interdependence," p. 48.

14. Ibid., p. 57.

15. Ibid., p. 56.

16. Jahangir Amuzegar, "The North-South Dialogue: From Conflict to Compromise," *Foreign Affairs* 54, no. 3 (April 1976): 552.

17. Ibid., p. 553.

18. The International Resources Bank (IRB) was proposed at the UNCTAD IV meeting in Nairobi, Kenya, in June 1976. It was designed to facilitate trilateral concession agreeements in which a consortium of entrepreneurs, the host-country government, and IRB would participate. The IRB was designed to minimize the element of political risk for the consortium and was to enable non-energy-rich less-developed countries to acquire the expertise necessary to develop their indigenous energy resources. The IRB was to provide a mechanism for financing buffer stocks of commodities subject to wide price fluctuations.

Chapter 4

1. Jacoby, *Multinational Oil*, p. 95.

2. Ibid., pp. 95-96.

3. Ibid., p. 96.

4. Ibid., pp. 96-117.

5. Blair, *The Control of Oil*, p. 211.

6. Ibid.

7. Ibid., p. 215.

8. Robert B. Krueger, *An Evaluation of the Options of the U.S. Government in its Relationship to U.S. Firms in International Petroleum Affairs*, prepared for the Federal Energy Administration, February 1975, Library of Congress, no. 75-906, p. 62.

9. For an objective analysis of company actions during the embargo, see *U.S. Oil Companies and the Arab Oil Embargo: The International Allocation of Constricted Supplies*; Report prepared by the Office of International Energy Affairs, Federal Energy Administration for the Subcommittee on Multinational Corporations, Senate Foreign Relations Committee (Washington, D.C.: Government Printing Office, January 1975), and *Report by the Commission on the Behavior of the Oil Companies in the Community during the Period from October 1973 to March 1974*, Commission of the European Communities, Brussels, December 10, 1975.

Chapter 5

1. For a discussion of these issues, see *Nuclear Power Issues and Choices*, Ford Foundation Report (Cambridge, Mass.: Ballinger Publishing Company, 1977), and

Energy: Global Prospects, 1985-2000, Report of the Workshop on Alternative Energy Strategy (New York: McGraw-Hill, 1977).

2. Unlike a conventional nuclear reactor in which, during the reaction, more fuel—uranium 235—is consumed than fissionable plutonium is produced, a breeder reactor, using plutonium fuel, actually generates more plutonium than it consumes. When a plutonium nucleus fissions after absorbing a fast neutron, it releases an average of three neutrons. One is needed to keep the reaction going and the other two can be captured in uranium 238 to form two new atoms of plutonium 239, thus creating more fuel than is consumed. See Ted Greenwood, George W. Rathjens, and Jack Ruina, "Nuclear Power and Weapons Proliferation," Adelphi Paper 130, International Institute of Strategic Studies, pp. 11-15.

Although a breeder cycle based on thorium 232/uranium 233 is possible and would reduce the need for world uranium supplies, most theoretical research has focused on the uranium/plutonium breeder cycle. Likewise, although uranium 233 is much less toxic than plutonium 239, uranium 233 is weapons-grade material and its large-scale introduction into the world community could complicate other nuclear weapons proliferation objectives. For a detailed discussion of different breeder-reactor technologies, see John Gray et al., *International Cooperation on Breeder Reactors* (New York: The Rockefeller Foundation, May 1978).

3. Theodore A. Wertime, "Is This the End of the World's Golden Age," *Washington Post*, April 24, 1977.

4. For an extended discussion of these issues, see Charles K. Ebinger, "The International Politics of Nuclear Energy," *Washington Paper*, Georgetown Center for Strategic and International Studies, Washington, D.C., Fall 1978.

5. For a thorough discussion of the components of the nuclear fuel cycle, see Greenwood et al., "Nuclear Power," pp. 11-15, and Gray, *International Cooperation*, chapt. 1.

6. For an elaboration of the issues raised by the development of nuclear commercial power as well as those of the manufacture of weapons-grade materials, see David J. Rose and Richard K. Lester, "Nuclear Power, Nuclear Weapons and International Stability," *Scientific American* 238, no. 4 (April 1978): 45-57.

7. For differing views on the sufficiency of worldwide uranium reserves, see *Nuclear Power Issues and Choices*, pp. 71-108; *Energy: Global Prospects, 1985- 2000*, pp. 202-6; Ranger Uranium Environmental Inquiry, First Report (Australian Government), pp. 57-72; and Ebinger, *The International Politics of Nuclear Energy*.

8. For a discussion of a market-sharing approach to reduce the threat of unbridled allied competition, see Abraham A. Ribicoff, "A Market Sharing Approach to the Nuclear Sales Problem," *Foreign Affairs* 54, no. 4 (July 1976): 763-87.

9. Gray, *International Cooperation*, chapts. 5 and 6.

10. It is currently uneconomical to recycle plutonium as fuel for light-water reactors; American Physical Society, "Report to the American Physical Society by the Study Panel on Nuclear Fuel Cycles and Waste Management," *Review of Modern Physics*, Supplement (January 1978).

11. Although the availability of uranium reserves is calculated on these prices, current uranium ore supplies are being marketed at between $27 and $40 per pound.

12. *Uranium: Resources, Production and Demand*; Joint Report by the OECD Nuclear Energy Agency and the International Atomic Energy Agency, December 1975; *Nuclear Power Issues and Choices*, chapt. 2; *U.S. Uranium Production Outlook* (Washington, D.C.: Energy Research and Development Administration, September 1976).

13. *Energy: Global Prospects*, p. 203.

14. See, for example, *Nuclear Power Issues and Choices*, chapts. 11-13.

15. See statement by Jerry McAfee, chairman of the board and chief executive officer, Gulf Oil Corporation, before the Subcommittee on Oversight and Investigations, House Interstate and Foreign Commerce Committee, June 16, 1977.

16. For background on formation of the London "Suppliers Club," see *Strategic Survey*, 1975, International Institute for Strategic Studies, p. 14.

17. For an elaboration of these relationships, see Ebinger, *The International Politics of Nuclear Energy*, chapt. 8.

Chapter 6

1. H. Gary Knight, "International Jurisdictional Issues Involving OTEC Installations," in *Ocean Thermal Energy Conversion: Legal, Political and Institutional Aspects*, ed. H. Gary Knight, J. D. Nyhart, and Robert E. Stein (Lexington, Mass.: D. C. Heath, 1977), pp. 45-73.

2. *Informal Composite Negotiating Text*, Third United Nations Conference on the Law of the Sea, U.N., Part V, Article 56(1)(a)(b), 1977.

3. Convention on the Continental Shelf, Article 2(1) [signed April 29, 1958; in force June 10, 1964] 15 U.S.T. 471 (1964).

4. Convention on the High Seas, Article 2 [signed April 9, 1958; in force September 30, 1962] 13 U.S.T. 2312.

5. For example, Lockheed Missiles and Space Company has developed one of the design concepts.

6. Knight, "International Jurisdictional Issues," pp. 67-68.

7. Robert E. Stein, "International Environmental Aspects," in *Ocean Thermal Energy Conversion: Legal, Political and Institutional Aspects*, p. 123.

8. Ibid., p. 124.

9. Ann L. Hollick, "International Political Implications of Ocean Thermal Energy Conversion Systems," in *Ocean Thermal Energy Conversion: Legal, Political and Institutional Aspects*, p. 85.

10. Ibid., p. 77.

11. For further information, see *Energy from the Ocean*, Congressional Research Service for the U.S. House of Representatives, Subcommittee on Advanced Energy Technologies and Energy Conservation Research, Development and Demonstrations, Committee on Science and Technology, 95th Cong., 2d sess., 1978; and Arthur W. Hagen, *Thermal Energy from the Sea* (Park Ridge, N.J.: Noyes Data Corporation, 1975); Jerome Kohl, ed., *Energy from the Oceans, Fact or Fantasy?*, Conference Proceedings (Raleigh, N.C.: North Carolina State University, January 27-28, 1976); and Adrian F. Richards, "Extracting Energy from the Oceans: A Review," *Marine Technology Society Journal* 10, no. 2 (February-March 1976): 5-24.

12. S. H. Salter, D. C. Jeffrey, and J. R. M. Taylor, "Wave Power—Nodding Duck Wave Energy Contractors," in *Energy from the Oceans, Fact or Fantasy?*, p. 5.

13. Richards, "Extracting Energy from the Oceans," p. 12.

14. E. L. Lawton, "Time and Tide," *Oceanus* 17 (Summer 1974).

15. William E. Heronemus, "Using Two Renewables," *Oceanus* 17 (Summer 1974): 24.

16. H. A. Wilcox, "The Ocean, Food and Energy Farm Project," Paper presented at the International Conference on Marine Technology Assessment, Monaco, October 29, 1975, cited in Richards, "Extracting Energy from the Oceans," p. 12.

17. For the Philippines' program, for example, see Albert Ravenholt, *Energy from Heat in the Earth* (Lebanon, N.H.: American Universities Field Staff, 1977).

18. *Energy from the Ocean*, p. 325.

Index